A CAUTIONARY LIFE

A Science Fiction Approach to Trauma Management

LYNNE CHRISTIE

ISBN
978-1-958122-56-3 (Paperback)
978-1-958122-55-6 (eBook)
978-1-958122-57-0 (Hardcover)

Dedication

This book is dedicated to my daughter and her family.

Table of Contents

Introduction

A Cautionary Life is not a biography, but it contains shades of my own life in it. It is a work of fiction but as I reread what I have written, it rings a bell of truth in me. It is intended to be of benefit to anyone who has suffered traumatic experiences in their life and are still dealing with the backwash. I found that journaling helped me in my own circumstances. The problem with journaling was that the real subject matter upsetting my life was too sensitive to commit to paper. I wrote volumes skirting all around the subject and avoiding the topics that hurt most. I overcame this by creating stories that helped me deal with the emotional pain. I found that it was easier to have an opinion about a story or someone else's story than it was to be logical about my own life's experiences. Gradually the things that hurt the worst became more tolerable, and I was able to address the most difficult topics directly. That is why this book exists. I wrote it to heal myself. I hope that you will be able to comment on the stories I have written and by doing so unleash your own traumatic experiences. Journaling does have a positive effect if you take the time to let it slowly release you. The traumatic events described in this book are very intense and not for the weak of heart. It is not intended to be read or used by children. It is graphic and direct. Finding your way through rampant interior pain is challenging, but the end result of release from buried torment improves life quality and is definitely worth the work it takes to gain the improved state.

Life is a dance of intricate balance. Good means little unless you compare it to bad. Being alive seems like a challenge instead of a gift until you face the death of a loved one. Bravery cannot exist until cowardice is conquered. You will not experience extreme happiness or good without extreme sadness or evil in existence. I realize that this sounds unfair. Why do ugly, harmful experiences happen to good people? I believe that we all choose before birth what experiences we will go through in our lifetime. This life is about gathering information. We need to use this information to decide whether we want to serve others or if we want to serve the creator within. This sounds like a simple decision, but it really impacts our quality of life. To serve the one God within, we will be giving up love. Love has no reason for existing unless there is someone we love, and love does not control others for self-gain. On the other hand serving others more than we serve ourselves requires sacrifice. If you were never challenged, if you were never controlled by the loveless, or gifted others without expecting reward, how would you absolutely know which you would choose?

All of us do, at one time or another, make a wrong decision. These negative decisions can leave scars from which we need to heal. Other times we are caught in the jaws of evil and shaken almost to death. To say we need to heal from these experiences is understatement. Life is occasionally not fun, but you will live and you need to continue growing. The quality of your life will be determined by your devotion to the healing process.

So, we are all involved in a dance competition with opposites and extremes where clear boundaries have become blurred. Honesty, as a constant, left when modern advertising arrived. Compassion has been diminished by greed. Even the lines of acceptable social behavior have changed so many times that there are no concrete guidelines. A large part of life is just surviving the circumstance in which we find ourselves. We don't even know how we got there, and helping hands are absent. What I'm getting at is that we have to come to our own conclusions about what we believe is right and wrong. No matter what we are taught, we will run into discrepancies that we have to resolve and come to our own conclusions. So, I hope this little book helps you to work through your own painful experiences as you weave your way through shades of my cautionary life.

Childhood

I've heard childhood defined as, "The time period of life from which we spend the rest of our lives recovering." Maybe "maturing" is just coming to grips with childhood traumas and fears. Some people are over achievers who seem to flourish without effort. Others stumble and lazy their way through life, while some are angry and rebellious. No matter our attitude, we still grow, change and become. Bottom line, we all have to start with what we are given and make a go of it.

Those impressionable, tender years between zero and eighteen leave an undeniable mark upon us. For some, the scars run deep. The following story is from my own life. The grandfather I speak of is my mother's father. I still love him and my grandmother, but if they were still alive I would have many questions for them. Forgiveness is mandatory. Understanding why people who are basically good, can do terrible things is not information which is given to us. It must be chewed, pondered and finally resigned to the unknowable.

Chapter 1

Childhood

My mind turns in on itself as I reject sleep tarrying over regrets, words not spoken, help not received, and blame placed (justly or not). Like swirling tarnished silver my memories float endlessly in a black miasma that I can't stop or even slow.

I was a baby when it happened: toddling but not yet running. I spoke only a few words but I understood most of what I heard. My world was filled with loving family members. I was a child who was wanted. My parents were married five years before having me and they believed I was the answer to their prayers. My mother was a school teacher and my father was a war hero. As WWII came to an end, my father came home with the Navy Bronze Star for bravery and a purple heart, but along with the honors came binge drinking, nightmares and fears that he buried deep. He had trouble getting his act together and she was the picture of responsibility. She just couldn't understand why he didn't leave it all in the past and move on. Personally, I understand my father better now than I ever did while he was living. The past can be a nightmare that you are experiencing while you are wide awake. You can respond inappropriately because the past is influencing you and you don't even realize it. You can choose a mate or friends because they resemble people from your past; and you can, for sure, choose to do the wrong thing at the wrong time because you have unresolved issues with the past.

I am thankful every day that I was born into such a loving family because I don't think I could have made it to old age if they had been any less caring.

My parents decided that the big city might offer Dad more career opportunities, so they left me with my mother's parents while they tried to find employment and housing there.

After they had left on their trip, my grandmother's sister called and wanted her to attend some event, but "Please, no children." My best guess at the time period is somewhere between fall of 1951 and spring of 1952. My grandfather, a proud man, worked hard for a living as a tool and die man five days a week and was a minister of a small Christian congregation. Every Sunday he held two church services and also one on Wednesday night. He went out visiting at least one night a week and then there was a brother's meeting on Saturday evening to prepare for the Sunday services. My grandparents lived on a small farm into which they both put large investments of time and energy. They were responsible and hard working. He was devoted to his faith and his family and was respected in the community.

Barton and Gladys had four beautiful children; Martha, Maler, Mary and Marven. Their children were born over an eighteen year period so only the first two were close in age. The youngest was eleven and a half when I was born to the oldest of their children, Martha. Marven, the youngest, was still at home when I came to spend time with them as was Mary, the second youngest. Marven was athletic and enjoyed hunting and other outdoor sports. He idolized his big brother, Maler, and often spent time with him and his family. He was always going somewhere and was rarely at home. By the time I was born, Maler already had two children, one girl, Bonnie and one boy, Maler Junior. Marven must have been with Maler and Mary must have been with Annie, her best friend, because when Grandma left the house that fateful morning, I was left all alone with Grandpa. My crib, Marven's hand-me-down, was in the dining room near the bathroom door, but to the right of it, in front of a window that faced south and looked over Grandma's cutting flower garden. The dining room table was massive with five heavy turned legs. One at each corner and one in the middle. I remember thinking they were like decorated tree trunks and under the table was big enough to hold the whole world. It felt sheltered there and happiness lived in that wood. The wide edges of this table supported huge family dinners along with loud laughter and bold claims to eating fame and prowess.

Also in this room, central in the home, was a huge grate in the floor. The coal furnace was directly below the dining room and drove volumes of heat into the house through that grate. In the winter it was delicious to sit or stand on the grate as it warmed your toes and made the circle skirts billow round with the force of the air. On the north wall stood a seven foot tall china cabinet that held all of the treasures imaginable. The good dishes, special serving pieces, an ebony elephant with ivory tusks, bowls with hard candy in them sparkled in the light and there were glasses of every size, shape and color filling the cluttered shelves. Grandma would hold me near the sparkling glass of the cabinet and tell me all about the wondrous things inside. There was a small ornate miniature chest that came from the World's Fair and there were gifts from Aunt Charlotte, Aunt Bertie and Aunt Dorothy, her sisters. Grandma loved these bits of glass. I could tell by the way she talked about each piece. They held love and good memories for her.

There were three other pieces of furniture in the room. One was a drop leaf desk for paying bills and it had a dining room chair beside it. One other extra dining room chair sat beside the china cabinet. The west wall was shared with the kitchen. On this wall was a buffet cabinet also filled with dishes and the silver grate was directly in front of this piece of furniture. Grandma kept everyday serving pieces, silverware and table cloths in it. The last piece of furniture was a brown low affair. It was perfectly square with a seat made of canvas strapping that fit down in it forming a seat with table surface on three sides. It was decorated with nursery rhyme figures and the alphabet. It was called a "baby-tenda" and it sat in the arched doorway between the living and dining rooms. It was normally a safe and happy place for me to play when my Grandma was busy. For the most part I liked it, but after this day I would never willingly get in it again.

I was in the baby-tenda when Grandma left looking very dressed up in her pillbox hat with the netting on the front of it. On her dress coat she wore a purple orchid corsage, and she wore white gloves. She always wore white gloves when she went somewhere dressy because her hands were dry, cracked and sore looking. Sometimes they bled and left red stains on her gloves. When she got two or three pairs soiled she would put them in a white enameled metal bowl in the kitchen where they would soak in bleach to get the blood out. Grandma had a wringer washer in the basement which was considered quite the luxury. The clothes were dried on a clothes line on the south side of the house because that side was always sunny. In the winter they were dried on lines pulled across the basement. There was no such thing as air conditioning so the windows were wide open all summer long. I loved the smell of the laundry that came in on the breeze by my bed. It was a happy smell. All good things came from my grandmother. She was a stellar cook who put fresh bread in the oven once a week and three large meals on the table every day. Sundays were exceptional. She would brown four or five chickens in butter in three iron skillets and then, in a huge Dutch oven, she would layer the chicken with wedges of onion. She spiced it with sage, allspice, poultry seasoning and thyme. Those fragrances still take me back to that world that she created for us. There was always company on Sunday. Someone different was invited every weekend.

It was her patient, kind, never hurried but always busy ways that structured my days as an infant and I learned how a woman approached life in that time. My grandfather worked outside the home and on the small farm to provide a certain level of luxury in life, but my grandmother made our lives rich with her ways. I spent most of my first two and a half to three years with her, because almost all of my earliest memories are of her. She was always kind no matter how busy she was and her compassion for small people had no rival in the world. She was born, Gladys May Franklin, third to the youngest of twelve children and her family was very close. I wish I could have known her parents and siblings better, because to this day I feel very close to them in spirit.

Needless to say, whenever my grandmother was away it was a loss for me. I remember she hesitated before leaving. She felt guilty about going and leaving me at home, but Grandpa guffawed and said she was being silly. I would be just fine for one day without her. She gave him instructions about where food and milk was for me and he shooed her off. It's too bad that he didn't listen to her. I did not received one drop of anything to eat or drink for the entire time she was away. I was in the baby-tenda when she left. It was early morning before the sun was up. I couldn't imagine why she was so worried. Grandpa and I would sing songs and go for a ride on the tractor and he would bounce me on his knee and it would be just fine until she got back. I've been wrong many times in my life, but this day I was so wrong that my entire life changed because of this one day.

After she had gone he came over to me and put a few toys on the tray. I held up my arms for him to take me, but he shook his head no and said I would have to play by myself in the tenda for a while. Then he turned and walked out of the house. I watched the sun come up and the room turn light. I was uncomfortable with my legs dangling so I bounced with my toes barely touching the floor. I bounced until my legs grew chafed from the seat straps rubbing my legs. My diaper was wet and cold. I called to Papa but there was no answer. I cried myself to sleep and woke stiff and cold, but still no one came. I was hungry and I regretted throwing my toys off the tenda in a pique of anger. Now I had nothing to do. I looked at the nursery rhyme characters painted on the tenda tray. There was Jack and Jill, Little Miss Muffet, Jack jumping over the candlestick and Jack Sprat and his wife sitting at a table with an empty meat platter on it. I liked Jack and Jill the best because they were free and playing, but Jack Sprat and his wife had eaten all the food and that started me crying all over again. The hours passed and I cried myself to sleep at least two more times. The house grew dark again and I was still trapped in the baby-tenda and there was no sign of anyone. No sounds to give me hope that they even remembered me. My tummy hurt so badly and I was shivering with cold feeling completely abandoned.

I was so relieved when I heard the back door slam and the kitchen door open. Grandpa finally came and got me out of the prison in which I had been confined. He staggered toward me and he bumped my legs hard on the edge of the table as he pulled me out of my seat. We almost fell down. He smelled bad. I could smell oil, sweat and whiskey. He took off my soggy wet diaper but instead of cleaning me like Momma and Grandma, he picked me up and licked me between my legs. At first the warmth felt good, but his whiskers hurt and he sucked so hard on me that I cried out and squirmed to get away as I started crying all over again. My voice was hoarse from dryness and crying all day, but cry I did; becoming more hysterical all the time. He spit between my legs and rubbed hard and clumsily. Next came the thumb. His fingernails were dirty and his thumb nail was split making a sharp spike on the left side of the nail. He pushed that weapon into me over and over again until the blood ran wet down my stomach. The warmth felt good. He held me upside down by one leg as he undid his overalls and dropped them to the floor. He fumbled with his underwear and then grabbed me by both legs. My head shook up and down as he forced himself in to me. I screamed again and again between gasps for breath. I was completely hysterical. He was so drunk that he couldn't come so he tried to push himself into my little anus. I don't know if he succeeded or not because by then I was in and out of consciousness. I was having trouble breathing as my left lung had collapsed, so he threw the screaming, bloody, limp mess into the crib with the words, "Worthless". Unfortunately for me, I believed that worthless must be what I was to be treated so badly. Sometime later he came back and put a diaper on me probably to keep Grandma from knowing what he had done. I must have been unconscious because I don't remember this at all, yet I know it must have happened as I had a diaper on in the morning.

While I was unconscious or maybe asleep, I dreamed. I left my broken body behind and I was moving down a long corridor of white light. I came to a group of people who were all very happy to see me. Their expressions were loving and although no one's mouth moved with a voice, they were all talking to me at once. Two women sat in chairs at the front of the group and the men were gathered around behind them. The woman on my left was heavy set and wore a long black skirt and a long sleeved white blouse with a brooch at the front of the collar. She smiled at me and pulled me up on to her lap. She cuddled me and talked to me of a great many things that I no longer remember, but I know that she told me she loved me and that everything would be alright. I felt safe in her arms and they seemed to give me strength as they communicated with me. The woman to my right was my favorite right away. She had beautiful soft features. She was slightly built and her hair glowed and made soft curls around her face. She said her name was Emily and I remember her as being positively radiant with light.

She sang to me and told me little stories and after a long while she said that it was time for me to go back. She told me that there were a great many things I had to do once I got back. I told her Papa was bad. She smiled at me and said she knew that, but it would be okay anyway and that I needed to forgive him. I said that I already had, but I didn't want to go back. Could I, please, stay with them? They chuckled and smiled and said it wasn't time for that yet. There were things that only I could accomplish which needed doing. And so, they hugged and kissed me very soundly and sent me off back down the hall of light.

When I awoke, I could hear my grandparents talking in the kitchen. Even though the abuse had been terrible this is still the hardest part to write about.

Gladys said, "Bart, there's blood everywhere. What did you do?"

In a gruff voice my grandfather responded, "Oh, you know there's always blood after the first time. You make too much of this. It doesn't hurt them. They have to learn sometime. Might as well learn young."

"There's too much blood, Bart! You stupid old fool. You've injured her," there were tears and fear in her voice. "Look at your hands! Look at your nails! You've cut her to shreds. What could you have been thinking?"

He hesitated a moment and then in a softer voice, "I really don't remember what happened. I had a mite to drink, but you know me. I wouldn't hurt a flea much less one of the babies. I love my baby girls and I'm the best one with them to put them to sleep. I never did nothen to hurt that child."

I didn't think this conversation was going like I thought it should have. She wasn't angry with him at all just shocked and frightened. I thought he would at least be sorry for what he had done and be honest with her about it. I felt I needed to tell Grandma, somehow, that he could not be trusted. He was lying to her. He was bad and mean, and yet I knew at that moment that she would never believe the truth, even if I could manage to tell her.

I called out to her. My voice was small and hoarse, but I tried again until I heard the kitchen chairs moving.

I managed to stand before she got to me, but I was dizzy and weak. She got to me before I fell back down. She told Grandpa to put some water on to boil and she talked to me while she laid me back down to take off the diaper. It wouldn't come off. The blood had dried to my torn skin. She closed the diaper back up with gentle hands and teary eyes. She picked me up and she talked to me about the things in the china cabinet as she tried to fill time while water boiled. I told her, "Papa bad!" but she shushed me and said that wasn't so. A small part of me gave up. She would not believe me over him. Another part of me was angry and yet another felt used and worthless. Even my best defender didn't believe that Papa could be bad. She didn't want to hear it. She didn't want me saying it. So, is this what life is like? I was so sad, terrified and depressed. I could not just resign myself to accept that what that man had done was acceptable. A burning rage was kindled in me. Even now, as an adult, that fire still rages. The thought that any man could do such a thing to a baby and go unpunished is unthinkable.

Grandma's white enameled dish pan was used as a sits bath for me. Diaper and all went into the tepid water. It stung some, but the warm water was mostly soothing. Soon the water turned bright red and the diaper fell away. There was so much blood that Grandma told Grandpa to go boil more water for the bath tub. The water in the basin was growing cold and I had goosebumps by the time he came back with a large kettle of boiling water. Grandma fixed the water and put me back into it before she completely broke down in tears. She didn't want me to see her crying so she left the small bathroom leaving my Grandfather sitting on the toilet beside the tub. I started to cry at the very thought of being left alone with him. He reached over and picked me up out of the tub. I became hysterical immediately. I just knew that it was going to start all over again and I was beyond consoling.

Grandma came right back into the room and took me away from him. She was stern and angry. It was the only time I ever saw her angry with him. She called him an old fool and he looked surprised. I stopped crying as soon as she took me away from him, so she put me back in the water. He declared that he had never done anything to hurt a child in his life and he didn't understand why she was making such a fuss about a little crying and some blood. I felt sick to my stomach as I realized that she would not stop him. She always gave in to him. She always believed that he was right in the long run.

This traumatic episode became buried in my subconscious. I completely forgot that it happened, but my grandfather kept trying to convince my grandmother that he had been in the right by insisting on "putting me to sleep" at night. This consisted of him rubbing me until I climaxed over and over and fell into exhausted sleep. They all kept giving me back to him until I was about two years old. By then, I had enough strength to fight him and I could talk pretty well. Mom took me from him one evening while we were all watching TV together. She took me to change my diaper but she asked him why one of the diaper pins was gone. He said he didn't know. "It probably fell out." But, I think she knew right then that he could not be trusted. We moved out of the state within the next couple of months. I believe that it was easier for them to leave everyone they knew and loved and move two hundred fifty miles away than to confront my grandfather about his problem.

Years later, when I was in my early twenties, I told my Grandmother that I was going to divorce my first husband. I told her that I didn't want to have sex. I explained that I started crying and got very angry and just couldn't get past it. She said that I must not really love him because she had always wanted my grandfather in that way. She said she still missed him and wanted him. At that time he had been gone about ten years. Still to this day I can't imagine loving a man that much.

All of these memories came back to me bit by bit in dreams after my mother died. I've been working through all of it for about ten years now, and the rage has tempered some, but part of me believes that pedophiles should all be castrated. Another part of me realizes that problems like this are society based. Woman have to step up and protect their children. Fear has to be conquered and there has to be help available to victims. This kind of ancient cruel torture of children destroys the lives of the victims and tears apart families. If you can't remember what happened to you, you can't make peace with it, and you end up carrying around rage that can erupt at the most inconvenient times. Another part of the problem is that abused young girls come to enjoy the sexuality and will become part of the covering up because they do not realize the extent of the damage it is doing to them. This can be especially true if there is no pain or trauma involved in the abuse.

I spent my childhood yearning for the thrill of sexuality and not understanding why. I also found that I had a distracted mind with my thoughts always floating off toward sex. I discovered masturbation somewhere around the age of five and almost always put myself to sleep by climaxing. I have survived four failed marriages and I doubt very much that I will ever marry again. It's difficult enough to keep a "normal' marriage together without bringing all of this extra baggage into it as well. All of these negatives, as well as others, were a direct result of my grandfather's abuse of me. What parent or grandparent would wish this kind of emotional trauma on a loved one? Well, I guess, obviously a pedophile would.

Children are like spinning tops of information gathering. They focus long enough to understand what is going on and then they spin off in another direction after some new bit of data. A child that has been sexually awakened too early will be easily distracted which stops the normal learning process. They stop spinning and go, "Oh, what was that!?" They become thrill seekers. Often vain, distractible and always trying to find an experience as powerful as the ones they had in childhood. Of course, that is not possible as we become less physically sensitive as we age. So their lives become a search for something that doesn't exist. No adult will speak to them about it because, as a culture, we do not talk about what is healthy sexually. At 67 years young, I still don't know what is sexually healthy.

Another problem with abuse of any kind is that it keeps repeating in your mind. It's easy for someone who has not had violence rained down on them to say, "Isn't it time to get over it? Move on? You lived, so get going in a new direction." The words are good advice but self-righteous, and cutting. "Draw a line in the sand. Step over it and keep on going," they say. That would be neat and tidy, but it doesn't work that way. You can't stop your mind from repeating the events over and over. The trauma is more like water that seeks your weakest point. You can find yourself depressed for no reason. You can ache for sexual fulfillment but be completely incapable of interacting. You doubt yourself, feel worthless and still men rule the known world. I forgave my grandfather instinctually when I was just a baby. I was disappointed that he was not remorseful. He was not sorry for what he had done. He was a man and a man has rights. I was a girl child and expendable. We live in what we think of as a modern society, but we still embrace ugly attitudes of inequality and violence. We bow to the Gods of beauty instead of love and compassion.

As far as I know, he never recanted his actions. He paid no price for what he did. His punishment never came while mine has been unending. Pain, abandonment, feeling unloved, worthless and guilty are just a few of the gifts he gave me. To this day I still struggle with this kind of negativity. There was no line in the sand. There was no direction in which I could go to walk away from the damage he had done to my soul. It's permanent. The scars will always be there.

The part of me that approves of castration for this kind of crime believes that this would be medical justice not violence. If a man went out and shot a baby with a revolver, even given that the child would survive, would you give the gun back to the man and tell him not to do it again? I suppose that a perpetrator could be sentenced to life long counseling and wear a tracking device at all times, but I feel that even one mistake would be one mistake too many. I think that being safe, not sorry, is the better stand.

CHANGE OF HEART

Some time has passed, and I have learned. I was introduced to the Law of One and this has changed not only my thinking about this issue but all of my life.

The Law of One is the religion of the Universe outside of Earth. The belief is that all of existence emanates from a central sun from which all of the known universe is projected as a hologram. All is created by intelligent, infinite energy. We would say created by God. All is God. There are two ways of serving God:

1. Serve Others 51% of the time, or (Loves constantly)
2. Serve Yourself 95% (Eliminates love completely)

When you have achieved one of these two states of evolution, you are ready to graduate from third density. A density is a vibration, musical in nature which dictates the laws of physics and the nature of reality. Higher densities are more highly evolved. The higher the density the deeper the information envelop is.

Earth has been having a problem with third density for the last 75,000 years. Very few meet the requirements necessary to graduate to fourth density. After the first 25,000 years of this period there was no one ready to graduate. So the ancient and wise ones of our galaxy chose to create the veil of forgetting between the conscious and unconscious minds. This can also be considered a veil between (death) between space/time (our reality) and time/space (the 5th dimension). Now, to get this concept of "space/time" and "time/space" you need to think of a pair of pants where one leg has gone inside-out into the other leg. The pant legs are very close to each other. You can only see one leg but that doesn't mean that the one not visible is not there. Our loved ones are really very close to us and once the veil is lifted at the end of third density all will be set right again. We will no longer forget each incarnation. Reincarnation will still exist but we will be able to remember where our loved ones are going to rest and prepare for their next training session. Time/space (or heaven, right leg of the pants, the 5th dimension) is where our departed loved ones exist. We live in space/time or 4th dimension, or the left leg of the pair of pants. Now, all of this was done to serve as catalyst to promote our evolution. Then, through the teaching method of karma/dharma (which is: as you sew so shall you reap) and uncountable opportunities to practice (through reincarnation or life after life after life) we were supposed to evolve (grow in consciousness and expand our abilities).

In real reality there is no death. We are all cut from the same fabric and that fabric is God. God does not die. Energy cannot be destroyed. Therefore there is no such thing as death, only transformation. Now everyone in the Universe knows this except us because the negative polarity of this planet (5%) has very trickily kept us in the dark. We are easier to control when we are afraid, and they want us serving them.

Now that the backstory is set, let's get back to our pedophile. If we are all connected in God and we hurt someone else, then we are hurting ourselves and God. Therefore cutting off a pedophile's guilty parts is counterproductive to our own growth. It must be remembered that forgiveness does not condone the act, it frees us from an illness of hatred, contempt, guilt, or any of a million other negative emotions. Forgiveness frees us and leaves the perpetrator to his or her own karmic fate (and karma can be a real bitch). So, I have changed my mind. I have recanted. I forgive my grandfather unconditionally. I love him. I give him the freedom to keep his parts and continue on his journey. I hope it will be toward the light and love side in his interactions with others instead of the negative, selfish, self-serving side, but I realize that we all have free will.

Childhood

Emotional Wellbeing

The Mental Health Foundation defines emotional wellbeing as "A positive sense of wellbeing which enables an individual to be able to function in society and meet the demands of everyday life; people in good mental health have the ability to recover effectively from illness, change or misfortune."

A person who is anxious, angry, or fearful does not have a sense of wellbeing. They cannot explore freely or respond light heartedly, or enjoy a simple moment without their emotions being on edge.

Victims of abuse see life through a lens of fear. It takes great courage to face new situations. I guess my message here is to have courage. If you're going through hell, keep on going. If you keep going, someday you'll walk out of it.

People who are emotionally stable do not feel the need to hurt others to make themselves feel superior. They have a central core of beliefs that keeps them steady. This gives them a sense of self-worth. A core belief in themselves allows them to extend value to others. Far too many in our society are missing this framework from which to hang their lives. They are not stable emotionally. It is my personal opinion that coming to a consensus as a society about this kind of moral value system should be a high priority. We should agree on what respect we owe to ourselves and others. Then we should teach it, live it and defend it. There should be a baseline of principle upon which we all stand whether private citizen or corporate entity. The principles chosen should be capable of extending not just across country but eventually accepted world-wide and be capable of extending into the universe and be defended by Constitution and social expectation.

The Bill of Human Rights written by the United Nations in 1948 is very comprehensive, but I haven't heard about it being defended. If it is going to make a difference in our lives it has to have meaning for every human being. I doubt that there are many people who could quote from it. We all know what the Ten Commandments are, but are they still accepted? Many would argue that the Judeo/Christian morality is not modern enough.

Respect is what I see as a baseline. A person who respects himself is more likely to respect others and therefore be a functional family member and a useful citizen. The Cambridge English Dictionary says that respect is admiration for someone or something that you believe has good ideas or qualities. That sounds kind of wishy-washy to me. Respect is too important to be misunderstood. To me, respect is granting to myself, as well as to others, the right to the following:

1. The right to express ourselves without fear.
2. The right to live life without harassment or bullying.
3. The right to marry by choice.
4. The right to seek knowledge both earthly and spiritual.
5. The right to liberty and the pursuit of happiness.

Both the United States Constitutional Bill of Rights and the Universal Bill of Human Rights say these same things and more, but for some reason they are not held sacred. They should be. This is what we should teach and expect of all of our children and ourselves.

We have come to a point in time when "self-evident" is not the case anymore. Our forefathers were all deeply religious and had central core strength's well in place, but our society is different. We have replaced religious knowledge with spiritual experimentation and many have dismissed the need for either.

All fractional groups as well as individuals should be able to see the need to live a set of principles by which we can all prosper in harmony.

Chapter 2

Emotional Wellbeing

I t is always surprising to me how life can drone on and on with no change happening or expected and then suddenly one small incident changes everything.

Terra was dressed for a professional conference. Her ensemble, a navy blue skirt with embroidered red flowers, was topped by a soft red gauze blouse with a Peter Pan collar. She wore hose and a pair of soft red low pumps on her feet. Her long brown hair was pulled up into a high pony-tail while the length of the tail was turned under and fanned out in an arc to make a wide bun at the back of her head. It was bobby pinned securely across the bottom and sprayed in place so as not to move. Terra had celebrated her thirtieth birthday during the winter, but now it was spring, and it was abnormally cold for a late spring day. She wore her beige wool winter coat that closed with a belt that tied in the front. She was confident and she knew she looked good. The lobby was deserted but decorated with a display of art work. She was late but it would only take a few minutes more to check out the art show.

As she rounded the corner of the first display her line of vision could see a side door. The door opened and the handsomest man she had ever seen walked through it. His chiseled face was thoughtful. She could see the blue of his eyes even from a distance. Tall and slender he wore his gray suit well, looking dignified without appearing stuffy or old. She shuddered at first glimpse of him. Was it the drafty room or did she know him? She studied him briefly but he was looking right back at her and walking right toward her. He moved with the grace of an athlete. Embarrassed at having been caught slack jawed, she looked down and turned toward the display. She knew her face must be bright red. He came over to the first display cabinet, glanced over it and moved on. Eventually, they were looking at the last display case at the same time. Terra chose to overcome her embarrassment. "Do you like any of the art work?" She barely managed to sputter out.

"No," as he shook his head, "Not unless there's something of yours here."

"No. I wasn't drawn to anything either. The sculpture is some better, but really marginal."

He turned back to the cabinet he had missed and after a quick perusal, concurred with her evaluation of the work. "Are you here for this art show?" he queried.

"No, I'm late, I'm supposed to be in a lecture in the hall below."

He glanced at his watch, "I still have some time to kill. What is the topic?"

"Art History in the Elementary Classroom", she answered.

"You're a teacher then? How about we go listen to the lecture together?" His smile dazzled and there was a twinkle sparkling in his eye like in the cartoons.

She returned his smile and said that she would very much enjoy his company.

She felt a pang of conscience, but she convinced herself that there was no harm in sitting by a man in a full lecture hall. They sat on cold folding chairs at the back of the hall. The lecture was non-descript and it was almost a relief when it ended. They silently watched the crowd as it exited the room. The air was stuffy, and it had suddenly become too warm.

"So what do you do for a living," she queried.

"I have a small business called The Plumber's Choice and my day job is federal."

Terra smiled, "I didn't realize that there were a whole store's worth of plumbing fixtures. How did you get into that business? Wait, was it a family business?" He nodded that she was correct. She didn't comment on the "federal" part as she was turning that over in her mind. Federal

Savings and Loan? Federalists were from the Revolutionary War period, right? Federal, as in…National? What could he possibly do? She decided he was a banker. "I teach fifth grade language arts and history. Today I was looking for a way to improve the interest level in my history classes." She indicated the room in which they were sitting.

He smiled, "Good honest profession that. Do you like it?"

"I love working with children, but I don't like the politicking that goes on."

He shook his head in agreement. "I have the same problem."

She continued, "Too many teachers are people who have something to prove, or have a need to puff themselves up. Their first duty should be to the students not to themselves. All of that looking for fault, and one-upping is inappropriate. That's not the way I think it should be done."

"How do you think it should be done?" he asked.

She collected her thoughts and then responded, "Our education system is based on the factory model. Children are not manufactured so the basic model of a teacher standing in front of the class lecturing does not work very well for a high percentage of students. (Now, I'm talking elementary level here.) Modern scholarship has proven that children learn in direct proportion to the size of the group in which they are taught. A group of three or four is perfect and over five students in a group, drops the learning curve into the abyss. So, we should be decreasing our teacher student ratio, at least in the first three grades. We should be providing all-day kindergarten and preschool for anyone who wants it. And, get parents involved in the classroom itself as much as possible. Maybe provide economic incentives to businesses that give employees time to volunteer in their children's education. The system we are using is teaching to the average child. We lose the lowest kids because we can't get to them all with just one adult in the classroom, and the top kids are bored into amusing themselves."

"I was one of those looking for amusement," he shared.

"Part of the time, me too, but I learned well enough to get me through school. I think the children need nurturing and more social skills training than they are currently getting at the early elementary level. The curriculum is so demanding that they really don't have time to play like children. From Preschool to third grade all of the emphasis should be on reading and math. If they have learned a love for math and reading by the time they start fourth grade the sky is the limit for what they can accomplish. They can master organizational skills in fourth and start focused creative writing in.........oh, what am I doing but boring you to death with shop talk. I'm sorry."

"I've enjoyed listening to you. It's a perspective I've not heard," his voice like velvet. "Where did you go to school?" he questioned.

"My first year was at Malcom Johnson University, out of state. I had a little too much fun that year and got reeled back home like a fish on a line. Then I lived at home and did my undergraduate work at Trembau. My masters is from Dever University."

He chuckled to himself and said, "My undergraduate was at Dever and graduate work was at Marten University in Columbia. I wasn't much of a student back then. School felt like just one more thing that was expected of me. My heart wasn't in it."

She concurred. "I also have felt pressured to do well. Well, to be honest, I've felt pressure to be perfect all of my life."

"Exactly. My father became really ill while I was in my early teens. I had to step up and help with my brothers and sister as well as with household duties. It felt like a burden and shoes I couldn't fill but I did what I could."

Thoughtfully she said, "Accepting or escaping responsibility does seem to form a pattern...... I'm not sure." A quiet interlude passed. "I have a singing voice that people notice, and I've felt a responsibility to the talent, but it makes me nervous. I get so anxious before I perform. I chose to teach instead of using my voice to make a living. It may have been a mistake. It was just that so many people were pulling me to do this or that. I really felt pressured. It was as if I needed to start liking only classical music and I found myself in new and uncomfortable situations that were alien to me. Like learning to sing in different languages. It was just too much with no back up. I did win the All City Music Award from the Music Teachers Association for 'Most Talented' in high-school in 1969."

"It sounds like you might regret not pursuing music." He was looking at his hands as he talked.

"Yes, in a way I do. I just didn't know how to go about it if I didn't go one of the many routes that so many people were pushing at me. For example, my voice teacher wanted me to pursue opera. I didn't connect with the music all that well even though I could really sing it. I didn't understand the emotion that was in it. It seemed empty to me. I preferred popular music. I didn't apply myself like I should have, and she was really frustrated with me. She said that if I kept up my lazy ways I'd end up sounding like Drayhardt. Wait! What? And exactly what could possibly be wrong with that? It was the wrong thing to say to me and that was my last voice lesson."

"I know what you mean. My parents wanted me to go in to politics." He stared down as he spoke. "They want me to be the president."

"The president of what?" Linda queried.

"…of the United States of America. It has become a dream of mine as well."

Terra gave a short laugh. "Why would you want that job?"

He looked hurt for a moment.

Terra saw her mistake and quickly explained, "I've never met anyone before who actually wanted to do that.…you can't make any serious money. You can't please anyone and it's about as high stress as any job gets. Not to mention the fact that it costs a fortune to do it."

"I'm not very good at explaining things. I'm a strategist not a speaker. I have to write down my ideas to share them."

"Please try. I'd love to understand your motivation."

He hesitated then and slowly began in a quiet voice. "When I was in high school we lived in Neuton and my dad was friends with a man who ran for President. We were at his home the night the election results came in. He lost the election even though he was the best man for the job. I was thunderstruck by that defeat. The campaign strategy had been all wrong. He should have won that election.

I was president of the Young Republican's Club at school and from there on out I knew what I wanted to do."

"So, if you were to become president what would you try to accomplish?" asked Linda.

"I feel that I could make a difference. They need a chance.…more jobs.….they have so many needs.….structure is wrong sometimes.…can't fix them all, but still, doing something to help.…" As he grows quiet.

"So, a social commitment?"

"It will just get worse if I try again." He said almost sadly.

"Do you have a learning disability?" she asked him.

"I don't know."

"Well, a specialist might be able to give you some tips that would help you sort out your thoughts." She offered as encouragement.

I don't want to do that now. There's so much to do. We are moving to Washington in a couple of months."

"You're married?"

"Yes, we have three children."

"Oh…," Terra was devastated. Stay positive she told herself. "Well… I guess it's good to know that everything works."

He burst out in laughter.

She smiled too from ear to ear. "You know, some people do have problems. In that area."

With a big smile still on his face and a twinkle in his eye he asked her if she were married as well, noticing that she also was not wearing rings.

Her face went from playful to solemn. "Yes, I'm married with three children also, but they're not mine by birth. They are his from a previous marriage and I've been raising them for the last five years. It's not a happy marriage. I'm afraid of him. I've only stayed because I love the children. He's just so mean."

"I'm sorry it's a bad relationship. I have a very good relationship with my wife. She is my best friend and confidante. What does your husband do that is so mean?"

I was cooking dinner the other night and Joey, the oldest at sixteen, came up behind me and gave me a hug. He said thank you to me for everything I do for him and his brothers. It was so sweet and kind. Bill suddenly pops up from his lounge chair and crosses the kitchen like a lightning bolt. He grabs Joey by the back of his pants and shirt at the collar and literally throws him across the dining table and into the wall on the other side. The dinner dishes went crashing to the floor along with the boy. I just stood there, mouth gaping, my heart in my mouth. I should have left him right then. He's crazy. What could make him attack his own son like that? It had to be jealousy, but what was he thinking? Did he think that Joey was putting a move on me, or was he mad about something else and just took it out on Joey? No matter what he intended, he is definitely not stable. At any rate, it's his emotional instability and meanness that has ruined our relationship. I'm tired of the verbal abuse and constantly living in fear."

He looked at me and then down shaking his head. "I have a happy marriage and I love my children very much. My wife is my best friend, but we aren't close physically. We talk through everything. I guess I treat her like staff."

After consideration she said, "I think I understand. It's like when you try to picture your parents as lovers. It seems so unlikely. They are so capable and in charge. How could they have a weak moment? Well, in my parent's case, they definitely had sex once. After that I have my doubts. So you and your wife have had sex three times or so?"

"That pretty much sums it up," he quipped in a tight voice.

"Still, you are very fortunate to have such a strong friendship. Maybe the other will come along as time goes by. My favorite aunt and uncle began their marriage with a raging sex life, but over the years they have become really close friends. They don't seem to miss the other. I guess everybody has stuff in their lives that they just plain deal with."

How long have you and Bill been married?" he asked.

We've been married about six years. He's not my first husband though. I married the first time to a childhood friend. Our families are still close. I just didn't want him sexually. The divorce was completely my fault. I couldn't function in bed. I was terrified for no apparent reason and I finally just said I needed to call it quits. He was willing to keep me no matter what, but I was really a mess over it, so he filed for divorce.

I met Bill at a work related Christmas party, and by Valentine's Day we were married. I knew him about six weeks and I married him. Can you believe that?"

He smiled a crooked smile, "I can because my wife and I count our marriage beginning from our first time being together one week after we met. Of course, there was a reception for the family, but we really understood each other from the very beginning.

You said that Bill is mean. Didn't you realize there was something wrong with him during that six weeks?"

"I did," she admitted. "You're going to laugh at me, but I thought I was supposed to marry him. I prayed for a sign from God and I asked that if Bill was the man I was to marry that God would give me a clear sign.

Bill and I had gotten into a really big argument and he had left in a huff. I prayed asking that if Bill came back with an apology and a gift then I would know I should marry him. Bill is not the kind of man to give gifts or make apologies so I was really thinking that we were finished, but he came back contrite with a jade heart on a long chain. Very appropriate considering the marriage.

It's been hard, but I have learned a lot from him, both good and bad. It's just been difficult for me to admit that I've messed up once again and have to start all over for the second time. My heart breaks for the boys because they will have to go back and live with their mother and I'm going to miss them terribly. It's a sad thing to dissolve a marriage, even a bad one."

He was quiet for a while and then, "Why don't you come with us to the Capital? I'm sure we could find you a job. We always need help."

She smiled at him sweetly. "Thank you for that, but teaching is my career, not politics. Beside I've got a big mess to clean up out in the country. And, I have a feeling your wife would not approve of our friendship."

"Let's go get a bite to eat and we could talk about it. We could go to the Grand Chalet. You'd love it there. It's really cool. All the reporters hangout there."

"Wouldn't reporters think it odd for you to be out to lunch with another woman? I don't see that ending well for you."

"You can squirm all you want, but it's no use. You're mine now and always will be."

It made cold chills go down Terra's back at his declaration of ownership. It wasn't a statement made by a bully but a statement of observable fact. She wasn't sure when it had happened, but the chills confirmed that some sort of bond had been made. The whole morning seemed more like a dream than reality. They had shared secrets as if they had always known each other and had complete confidence in each other's trustworthiness. His declaration of love for his family made it clear where his loyalties lay, but still her heart knew he was right.

Terra stood reaching for her coat, "I really have to go. I'm going to have a hard time accounting for this time."

"Me too. I'm going to have some serious explaining to do." He reached for his coat and laid it over his arm.

Terra, unexplainably felt tears welling in her eyes, "I hope all your dreams come true," she said as she extended her hand to him in farewell. She leaned forward and up to kiss him on the cheek. She must have caught him off guard or in movement himself because their mouths brushed each other. There was a tingling that made her catch a quick breath. She wanted that kiss. She leaned back into him trying to catch his lower lip between hers. He responded with a cross between play and adventure. As they separated a string of spit still connected their mouths. Embarrassment and need hung in the air.

"If they do, you'll be there."

Emotional Wellbeing

READER'S COMMENTS

Physical Wellbeing

This is the most obvious category. The problem here is that I'm not an expert in this area. I am still struggling with my own physical wellbeing. I am a diabetic who has not been able to find a medication to which I do not have a bad reaction. Restricting food and exercise are the only tools I have to use, and so far these two tools have not brought my blood sugar down to where I want it. Perhaps I'm wrongly critical of fast food life, but really I don't think so. I did not believe that eating fast food would damage me. I believed that I was in a hurry and that all nutrition was about the same as long as you didn't over eat and you took a multiple vitamin. Definitely, I was wrong about that. I think the real snake in the grass here are simple metabolizing carbs. They taste good and it is way too easy to eat too many of them.

"You don't have anything if you don't have your health."

I've heard this idea all my life with just the wording changed, and it remains true. I believe that very few people in this American culture would claim to be at their ideal weight. Yet diet after diet fails. I've come to the conclusion that:

Fast Metabolizing Carbs + Cell Phone Face = FAT

You probably need to change your diet to eliminate or at least reduce your intake of simple carbohydrate and increase your movement. I'm sure you have heard that before. You CAN change your body. Our bodies are constantly changing. It is slow change, but change none the less. It is this "change ability" of which we need to take advantage.

It is also necessary to feel good about ourselves. We need to love ourselves, forgive ourselves, and celebrate life from whatever circumstances we are in. You might be surprised how much easier it is reducing weight when you are being kind to yourself and are capable of focusing on what you are putting into your mouth instead of being miserable about what you can't have.

Personally, I'm a nervous eater. I have to feel centered and loved to have the strength to say no to simple carbs. It's a sacrifice for me to leave them alone. What I'm getting at here is that we all have our struggles with the food issue, but you can't give up and just give in to these cravings. You need to have a plan and work it. Live a healthy life or you will be struggling with diabetes like I am. Choosing a healthy diet is far easier than being chopped up bit by bit because your body won't heal any more.

I have one other area of weakness. I make excuses as to why I can't exercise. "It's raining. I'm not going out in that." or "It's too cold, or too hot, or too muggy, or…there's a dog that barks at me, or I feel like someone is watching me, or I can't go the long way because I'll have to pass that certain house…" Come on. Really? Just walk. Do it. Don't let your fear rule your decision making. I've made a few bad decisions in my life, but the worst decision I ever made was to stop exercising every day.

Chapter 3

Physical Wellbeing

Terra was forty two years old and in good physical condition. She swam almost every day. She loved the way it made her feel. The smell of chlorine in the water, the shock of the cold water when she first got in and the jolt of adrenaline was addicting.

She had begun swimming four years ago when she had begun her divorce. She needed relief from tension as the stress and anxiety from the divorce experience had given her colitis. The exercise helped with the cramping and took her mind off of the emotional pain.

This particular day, the parking lot was unusually empty. Terra asked the lifeguard if it was alright to swim. She thought that she might have missed the memo to not show up today. There was one other woman swimming, and the guard, although she had never seen him before, affirmed it was okay to swim.

She decided to make the most of the available space. It was liberating to have so much room to move. Normally, they all had to swim two and three to a lane to fit everyone into the pool. She did exercises in the water for her stomach, legs and back with a few stretches tossed in to limber up. The water always felt cold when she first got in, but by the time she was on her eighth lap it felt warm and comforting. Terra extended each arm as it crossed her ear and felt the sweet strength as each pulled her through the water. Her feet only needed to flutter to keep her horizontal on the surface of the water. She swam faster and faster as her body warmed to its effort.

She could have kept swimming forever. There was no sweat or exhaustion just her own strength and sense of achievement.

Summer was almost here and for a teacher that was golden news. Prepping for the coming new term was her day job during the summer, but her schedule was more relaxed and that made it possible to swim for longer periods of time. She was reviewing her plans for the day as she walked to the showers.

Every four to six months she had to buy a new swimsuit as the chlorine wore them threadbare. Because of this, she took her time to carefully rinse out the suit to make it last as long as possible. She hung the suit on a wall hook to drip off and began to shampoo the chlorine from her long brown hair. The warm water felt relaxing and comforting as it soothed her tired muscles. It was a healthy kind of tired.

As she rinsed the conditioner from her hair she leaned back against the back wall of the shower. She let the water run over her head on to her face. As sudden as ice in winter, her feet flew from beneath her. She put out her left hand to keep from going down so fast. She felt blinding pain as something snapped in her left wrist. She looked to her wrist as she slid toward the opposite wall. Her hand was bent at a ninety degree angle and she could not will it to straighten out. Using her right hand, she pulled the all wrong left hand straight. A wave of nausea washed over her as the mind-numbing pain crushed all other senses. She made her way into the dressing room breathing hard and trying to keep her shaking legs under her. Pulling first one side and then the other her underpants and jeans got mostly on, but could not be zipped one handed. The other woman who had been swimming came to her rescue. The woman was a stranger but she helped Terra anyway. The long sleeved blouse was only capable of being worn on one arm and just wrapped around the shoulder of the damaged arm. The blouse would not close so Terra held it together with her good hand. The Good Samaritan drove Terra to an immediate care facility where a young female doctor gave her a place to rest.

The cool sheets were a relief from the heat of stress. Terra closed her eyes for a moment. She abruptly opened them again as she heard giggling and raucous comments not far from her. She was amazed to see three grown men dressed in rescue uniforms who were pointing at her. In her shocked state she could not imagine that they were actually laughing at her pain. How could they be so crude? There must be some other explanation. The young doctor had noticed their behavior as well and she came to close the curtain around the bed. Terra heard the doctor run them off, but the uneasy feeling in her stomach made her wonder what kind of place this was that had fourteen year old boys pretending to be rescue personnel. She could not imagine what possessed them.

The wrist healed nicely but her fear of a slippery floor never did. Terra gave up swimming. She always said she would go back, but she never did.

Physical Wellbeing

Chapter 4

Spiritual Wellbeing

My explanation of spiritual is the part of human existence that gives connection between the soul and the eternal. Spirit can be positive or negative, good or evil, or any combination of these opposing elements. To be spiritual is to feel a connection to the profound: positive or negative.

Soul and spirit are separate and different. Your soul is the part of you that has become, over who knows how many lifetimes, who you are as a person. It is also the part which is eternal and will continue on when your body dies. We are all aliens, of a sort, inhabiting an animated physical form. The body is the conveyor of the soul. Spirit teaches and writes upon that soul. We reach maturity through our experiences and the decisions we come to embrace. We cycle again and again until we mature to the point of seeing another way of being. We put aside the need for a physical presence and we "ascend" to accomplish other tasks.

In scripture there is mention of the return of Christ Jesus (the ultimate alien/human crossbreed). Christ ascended into heaven and went to sit at the right hand of his father. I'm pretty sure that had to involve space travel of some kind. And if he is going to return, again that will involve space travel to come from wherever in the universe he is, back to Earth again. That makes Christ alien. He is an 'off-worlder', and it takes alien technology to make this prophecy possible.

There are many humans who swear fidelity to one religious order or another, but what they are really doing is connecting with a spiritual energy that they agree with or to which they feel affinity. Far too many religious organizations claim to serve positive elements, but they, in fact, are serving cruel and misleading energies.

I believe in an infinite, intelligent energy which has created this universe. This intelligence is power, is mathematical and has used geometric patterning (sacred geometry) in all matter of the universe. On one extreme this power is in flux; growing, expanding, becoming. And at the other extreme, it is coalescing and coming to its conclusion. It has allowed all living things to mature in their own ways. This maturity leads living beings toward one extreme or the other. The ultimate question here being which is stronger; love or hate.

We all, whether or not we believe in it, are leaning toward one pole or the other. In this universe all ingredients are being sorted. In this third density creation the pathways are being established to worship God in one of two ways. We have free will to choose. Our training comes from adversity and injustice, and it is up to us to commit ourselves to one path or the other.

Our current universe is teeming with life and much of it was born earlier than Earth. There are intelligent races of life who have technologies which far exceed anything of which we have dreamed. We are all connected to the same battle. There are angels that lean and lead in both directions. Ultimately we all will have to choose who we want to have as our neighbors for all eternity. Will we choose for peace, love, joy, creativity, hope and courage or will we choose the opposite of each of these? Every small decision we make leads us toward one pole or another. Life from all around the galaxy will come to Earth for Armageddon. I'm rooting for all things positive and good to win that confrontation. Glory be to the one God.

Spiritual Wellbeing

Green glowing eyes radiated hatred as the restive monster paced the worn flagstones of his warren. His reptilian form showed the signs of needing resurrection once again. He would go to Armageddon in a new young body. The pealing places itched and oozed. When no one was watching, he picked until whole sheets of flesh fell off. He would miss that about this body.

He paced in even methodical steps as he pondered. If those puny humans thought to ignore him, they would find him an unforgiving god. The One God's plan had allowed him this war and he planned to take as many human forms into death as he could. All of his people from his earlier sending, their whole civilization, were gone now, but once he got there he would resurrect as many as he could find. He brought an army with him. His floating Zygot city carried millions, but he would add to his army and swarm over Earth.

He had kept a garrison on Earth since the beginning of mankind. They kept to themselves underground or in the water cities for the most part but kept an eye on things. He wasn't the only demi to have interest in Earth. Other beings watched as well. All would be gathering now who were looking out for their own interests. They would snatch what they could, but when they saw his fighting force they would kneel before his might. He would enjoy this victory.

Far, far, away, in another sector of the Universe, Athael was doing her own insubstantial pacing. She had deeply hurt Range with her explanation of why he was not ready for Earth yet. Her tirade, she feared, was colored by her growing jealousy of his relationship with Ponn, another trainer. She didn't want to admit that her student had become so important to her. She was so conflicted. She wanted him. She wanted to wear a body again so she could have him in an Earthly way. Athael was, uncharacteristically, out of control. Her duty was to prepare him for the last battle, not to corral his soul all to herself. During their last conversation she had done nothing but hurt his pride and push him away.

He was so accomplished. Athael had kept Range busy with one military campaign after another. He needed to be the best strategic general the Universe had ever seen. Life after life she had guided him. He was strong, charismatic and good. His campaign against the Dymar had been brilliant. Range was the stuff of legends and he even carried himself with humility. He was young and had not yet been schooled in how subversive the Demi of Marl could be. He did not understand warfare that was not direct and she had made a mess of things trying to explain it. He had turned red in the face and stomped off in a fury. As he left, she heard him say, "We'll see about that."

Why couldn't she just tell him how she felt? Why did she hide her emotions from him and hurt him on top of deceiving him? She was so angry with herself. Athael would need to apologize, but she would wait until she was centered and calm. Range was normally a level headed soul. Surely he wouldn't do anything rash. The presight globe had shown incidents on Earth where ridicule and belittling was going to be a tool used to lessen confidence in Earth's rulers. Range really needed training in how to deal with this kind of warfare. His pride was up and she feared he would not listen. She would let it rest. Give him time to cool off. Then she would try again, bringing a better set of tools to her presentation.

The garden in which Athael worked was a magical and beautiful place to her. The colors of creation swirled forming insubstantial sculptures that slowly flowed into new forms. The place where she and Range had argued looked muddied to her. She steadied herself as she let the calming pinks, yellows and blues roll over her form. The mists soaked through the worries of her job and soothed her, caressing her as they danced and moved about. She had been growing souls in this garden for eons, and Range was her finest student.

Ponn was another grower in the garden who did a similar job to Athael's. They were friends, but Ponn had recently declared her feelings for Range. Ponn had gone so far as to prepare a program for him for his next Earth sending. Range had not yet expressed how he felt about her plan so Athael had not worried overly about it. Ponn even had plans to be his wife in that lifetime. This was one of the many reasons why Athael had not declared her own emotions honestly. She did not want to be caught in a competition that would create negative feelings. Athael had, however, promised Range that she would go with him the next time he was sent to Earth. They hadn't discussed what capacity she would fill but she thought it might have been implied. That was as close as she had come to telling him how she felt.

Time meant nothing in this garden, or for that matter, anywhere on the Mountain. Time only mattered if it was in reference to Earth or another planet. The garden was a nebula. It was insubstantial yet teamed with life and organized activity. The One God resided strongly here and all was God. The enlightened lived here on many different levels. All who resided and worked here did so without a physical body. They would receive new eternal bodies at the Great Resurrection. Soul, the eternal material, coalesces here and becomes cogent, but not flesh. Athael had worked here thousands of Earth years and had lived countless lives on Earth in the great cycle before that. To her, life on Earth was a difficult business; whereas, her work on the Mountain, was heavenly. She knew that Range did not want to go back again, any more than she did. Leona had promised Range that she would be there to help him complete his mission, and she would keep her word.

After Range had walked away from Athael, he had raged inside himself for way longer than necessary. He loved her so much. Really, as far as companions went, he was not even interested in spending time with anyone else and he believed that she felt the same way. What infuriated him was the fact that she still found him lacking. He felt ready, confident, and strong enough to battle anything and win. Ponn had told him she thought he was ready. She was such a good friend.

He knew he could buffer hatred, build fires of compassion and make a path for humanity to follow to a more peaceful and compromising environment. He could provide guidance toward a strong sub-structure and help humanity strengthen relationships. He could do this on his own. He didn't really need Ponn to get the job done. He could use her program, but go alone. Athael really didn't want to go back to Earth again and he could save her that much. His decision was made. He would go alone, complete his mission and return home to Athael victorious.

Athael searched for Range everywhere. She was ready to make amends and she was raw already from assuming that he must be somewhere with Ponn. However, as she was heading back to the garden she ran into Ponn. They both spoke at the same time. "Where is Range? I thought he was with you," rang the chorus.

Quickly both ascended to the Gate Keeper. "Monspora, have you seen Range?" questioned Athael.

"Yes," responded Monspora. "He took a small launch and I saw a wormhole manifest. Having been distracted, I assume he took it."

Athael was struck with despair and Ponn didn't look much better.

"The next viewing is coming up soon, but you should not go during Timony's rule. He has declared for the Demi of Marl, and you know how he hates you. He'll take great pleasure in making your life a misery," Monspora reminded Athael.

"If I don't go now, I'll be breaking a promise that I made to Range. He must be operating under Ponn's program," said Athael thinking aloud.

Ponn quickly responded, "If Range is using my program then I'll need to go to help him. You don't need to go, Athael. You're not written in!"

"Don't worry, Ponn. I'll write my own program, although you may have to share." She said with an uneasy grin.

Ponn stormed off toward the hanger.

Monspora shook his head slowly. "I still think this is a big mistake for you to go, Athael."

Athael looked up into his kindly, concerned face. "I have expected others to face terrible circumstances in the cycle. I have asked them to endure incredible hardship. How cowardly would it be for me to refuse this? And, how would Range ever believe that I cared about him if I don't keep my word."

"But," spoke Monspora in a concerned and serious voice, "You may not be capable of pulling away from the cycle. Timony could even turn your soul to the dark."

Athael looked downcast for a flickering moment. "I will persevere, Monspora, and great good will come of it. Send me positive energy as often as you can, old friend."

"It shall be done. The grace of the One God go with you," intoned Monspora.

Spiritual Wellbeing

Relationships

Sailing Companion
Slack of sheet,
Sack of care,
Self-righteous,
Know it all, and Wrong.
I've been all these things and I see them in others.
Shame that I make others wear my weaknesses.
Sail of pride,
Generous always,
Compassionate,
Full of faith, and Brave.
I've been all these things too and I like to see them in others.
These don't seem as heavy with emotion, easy, no challenge.
Purse full of idealism.
Fairness flagged.
Who would you choose as companion?
YOU?
Red eyes, heart broke, betrayed, abandoned, and deceived
Curling up tight around one's self
Slowly sl ow l y
S l o w l y
Time passes, sleep.

A fragrance
Consciousness returns
The pain is sore now, not unbearable.
One eye opens.

Should two?
What is worth looking at?

You.

Hardship Mandatory / Suffering Optional

How to let it go… that is the trick.
When your stomach aches with nervous fear
Nothing is going right.
Pain that runs through your bowels and tenses your back
Like a bow readied.
Let it go.
Think of one thing right; a child's smile.
Are you doubled over from each wrenching blow?
Do your ears ring in jangling counter point to the torturer's attentions.
Breathe… for you can.
Think of two things right; a nice day of sunshine, a favorite meal.
That's better. The tension recedes.
You can think more clearly now.
Let it go.
You can't change what tries to torture you.
You can't make it stop. You are going to face it.
Then let it go. The sooner the better.
Hardship mandatory, suffering optional.

Simply put, relationships either make or break our lives. I cannot express clearly enough how critical others are in how we view ourselves, our lives, our achievements and our futures. Some relationships we need to run from, others make life's miseries endurable and others bring us sublime experiences. We all know these things, but occasionally we get caught up on a wave of emotion or lust and forget to watch for red flags. In the spirit of offering good advice, I'm reproducing a list of characteristics of a potential batterer from Ann Landers newspaper column dated April 1, 1995. I sure wish I would have had this list in 1975 instead of 1995. Speaking from personal experience, there is truth in these warning signs.

1.   Controlling behavior:

The most widely used statements for behavior restriction are "I know what's best for you" and "I know what you want (or need)." The reality is that no one knows what is best for another adult. Unless I tell you what I want or need, how could you possible know?

2. Blames others for his problems:

Common statements are "Look what you made me do" and "If you hadn't done that, none of this would have happened." When someone does something to you, you must decide how you will react. When Mr. Wrong and I had arguments, he would smother me with a pillow or choke me to get me to be quiet. I had to keep reminding myself that it was not my fault that we disagreed. It was not my fault that he lost control of himself. I am responsible for my behavior, and he is responsible for his.

3. Playful use of force in sex:

This usually includes demands. Mr. Wrong used to say that sex was one of my "wifely duties". There is no law requiring a woman to have sex if she doesn't want to. Participating in sex is not one of the wedding vows. Sex must always be by mutual agreement. Forced sex is called rape.

4. Past history of battering:

Excuses include the classic, "If you hadn't provoked me…." The truth is that he chose to hit, push, kick, slap or punch you. You did not ask him to do this. Most importantly, if he hit you once, he will most certainly hit you again. Count on it.

5. Verbal abuse:

This was Mr. Wrong's specialty. If someone deliberately hurts your feelings by word or deed, it is abuse, even if it is as simple as "You look fat in that outfit."

6. Threats of violence:

Threats are wrong because they are almost always precursors to the deed. If he threatens you, leave him before he does it.

7. Use of force in an argument:

Most women feel, as I did, that if they haven't been hit, they have not been physically abused. Restraining someone is also physical abuse. Pushing and shoving are physical abuse. Any time he lays a hand on you in anger, he is abusing you.

Ann, please tell your readers that abuse and battery take a toll on one's physical, emotional and spiritual energy. It is simple to decide that you no longer want to be abused. It is easy to say no. We say this word all the time. Unfortunately, we find it especially difficult to say no to those we love and those we fear.

The solution is to practice Zero Toleration. If you feel threatened or have been physically violated in any way, get out of the relationship. Tell yourself, "I don't deserve this, and I'm not going to take it anymore."

I thought it would be very difficult to leave Mr. Wrong. I worried about what others would say. I worried that he would retaliate if I left him. I loved this man, and part of me always will. However, I don't love anyone more than I love my life, health and sanity. All you have to do is decide that today you will be free. You can then be sure that tomorrow will be better. – Shreveport, LA.

Stop worrying! The bridges you build before you come to them are almost always over rivers that don't exist.

May I add an AMEN to all of that!

Chapter 5

Relationships

STORY

It was a gray cold May morning. Storm clouds moved and roiled as elephants changed to snakes in the sky. The wild wind moved the bluish purple of the Cantalorian grain growing in Milo's field. The grassy sea it made waved and swirled like water. The brisk air smelled of Milo's self-satisfaction and greed. He smiled to himself. The negotiation had been extra effort, but it would be well worth the risk. The grain, when harvested, would bring a huge profit. One more pass of KB27 and he would be a wealthy man in his early thirties. He stretched his torso and rolled his shoulders back as he remembered the day he had first noticed the Cantalorian merchant who had sold him the grain. He was a whiny sort who had been looking for a buyer for his seed grain at the town market. He stood out because is clothing was bright and his manner naive. Milo had watched for quite a while before approaching the stranger. The whiny merchant had a land rover that was filled with bags and bags of very expensive seed. Milo would normally not be able to afford seed so valuable. Cantalorian seed only needed to be planted once every ten years and the crop was so hearty that each year it was grown the harvest doubled. The merchant had not understood the value of that particular seed on Kril Birk. Milo had bid the man follow him home where an agreement could be reached while enjoying the comforts of home hospitality. As Milo remembered, his expression was not pleasant. He was arrogant, smug and superior. It filled him with satisfaction to remember how troubled the man had been when all four tires of his land rover had gone flat at the same time. Milo had offered to take the precious seed off his hands for a ridiculously low price. He had told the merchant that with luck and Milo's help he might make it back to civilization. So Milo had relieved the merchant of his merchandize, tightened the screws he had put in the tires, and aired them all back up again. The man might survive if he stayed to the more gentle terrain. The man had sorrowfully driven away.

Now what Milo hadn't discovered was the merchant's name. If he had known that he was dealing with Michael, he might had dealt differently with him. Michael most definitely did know the value of his product as it had been purchased with an eye to tempt. The immortal had traveled through five small villages before running into Milo. As far as Michael could tell, the majority of this planet's inhabitants were honest, kind and hardworking, but there seemed to be a secret underbelly that deserved his further investigation. Milo was small game but he might lead Michael to the strong negative pulse he had felt when first landing on Kril Birk. Petty crime had to be sanctioned to be arrogant and assured of no reprisals. This negative pulse had to be found.

KB1-10 had passed and Milo felt things had been going very well. He had planted only half of the precious seed just in case there was something wrong with it. He figured that he could always turn a profit selling the remaining seed to a mark if it turned out he wasn't happy with it for some reason. The crop was doing better than expected and he had managed to hire a passing vagabond who worked for room and board. Milo's wife was very pleased to have help with the chores. Wyette had not asked one thing of Milo since the old man had arrived two months ago. She did not even complain that her husband wanted to stay two extra days in Northfield this next trip.

Milo had a small farm tractor, but he feared that it was not enough to handle the harvest that was expected. He had purchased a burned-out, much larger tractor for a song and he was in the process of rebuilding the monster's engine. He believed that he was close to having it running. He needed to test it to see if the engine was still locked up or if it would turn over. Then he could purchase whatever parts he still needed while he was in Northfield.

Milo had set the hired man, Tom, to repairing the fence around the south field. Milo figured he could keep an eye on him that way as he worked on getting the H250 running. Wyette would drive the H250 and he would pull it with the D25 to jump start the beast. The wind had picked up speed and a gnawing edge of cold was biting by the time Milo got Wyette sitting like a doll on the huge frame of the H250.

Wyette did not trust Milo, and she almost believed that he was trying to kill her for the insurance money. The smell of her fear wafted on the breeze and dampened her clothing despite the cold chill. She set her will to survive. Somehow, she would persevere. Milo would not win that easily.

The big tractor began slowly moving as the smaller D25 struggled and strained to move forward. Slowly, ever so slowly the speed increased. Wyette had to hold on for dear life as she balanced against the seat with her backside, held down the stiff clutch with her left foot and formed the third leg of a tripod with her right foot for balance. The uneven ground of the south field that Milo had turned into was causing Wyette serious problems. She would fly up into the air and come down with every uneven ridge of earth. Her legs began to tremble and then shake with strain. She called to Milo to stop but all he did was turn, smile and accelerate. Wyette was thrown about like a rag doll in an angry child's hands. Somehow, she managed to hold on but her hope came from the sight of Tom running very fast across the field toward them. He was yelling to her to release the clutch. It seemed strange to her that a mature man could run that fast. On her next bounce skyward, she released the clutch and the metal beast lurched but it did not come to life. The stop was sudden and jarring throwing her hard into the steering wheel. Milo, half crazed kept trying to pull. He buried the tires of the smaller tractor in the field. There was a loud screeching pop as the large tractor pulled itself in half. The front part lurched forward and the back half, still holding Wyette, was left in the dirt. Milo's frustrated furious scream seemed to hang in the air. He jumped from the buried tractor with murder in his eyes. Wyette knew how bad his temper was and she knew that she had to escape somewhere until he calmed down. She decided to jump from the tractor on the other side. Her jump was to close and her foot caught the edge of the fender. She fell hard with the wind knocked out of her. She could not rise quickly. This happened just as Milo reached the carcass of his broken project.

Tom arrived just behind Milo not a moment later. Tom said in a quiet voice, not the least out of breath, "Milo, you should finish packing for your trip. I'll take care of this."

Milo's impulse was to hit the old man, but he never got the chance. A shimmering net of twinkling light fell over him and he knew no more. Tom simply and quickly wadded up the sparkles to a dime sized sphere and dropped it into a pouch he kept at his belt. He then rushed around the tractor to help Wyette to her feet. She was trembling and very disoriented to see Tom instead of Milo. Tom reassured her that everything was alright as he carried her to the old dilapidated barn. The straw bales were dusty, but the stack was still sturdy. "Milo has gone to finish packing for his trip," soothed Tom. "He will be gone soon and he will not hurt you." Wyette could not believe for a moment that Milo was not an immediate threat, but when she looked into Tom's eyes, she knew he was speaking truth. Tom encouraged her to rest in the barn while he did something about the tractors. He touched her temples as she reclined and she went immediately to sleep.

Tom went quickly around to where the two dead tractors were littering the field. He put a contraction field around both tractors and compressed them to the size of very heavy toys and discarded them in a weed patch. He then ran to his cabin and found an artificial intelligence unit that he had set back for just such an emergency. He was programing it as he returned to the barn. At a voice command from him, there was a new tractor sitting in the barn. The AIU would now become a tractor for the rest of its life instead of a land rover or a hover craft or whatever Michael wanted or needed it to be. Yes, he was Michael, merchant, old Tom or anyone else he wanted to be. Michael was a shape shifter, an Aberian, and disguises came naturally to him. He could change physically and he could also influence the perceptions of any life form that was in his presence. Many considered his kind to be magic but Michael didn't worry about it one way or the other. He had a reputation, but to him that was the same as gossip. All he did was his job, and he did it well. He had been sent by the Counsel to search this portion of the galaxy for signs of any evil seeding. The Compromise of 3552 CE forbade any further expansion by the Evil One, but it would not be like him to accept this judgement without recourse. Michael had been chosen for this job because he was Aberian and as close to invincible as any force for good could be. This quadrant of the Quarian Galaxy should have been free from the ugly ten, but there had been rumors, so Michael had been dispatched. He was sanctioned to discipline for any of the ugly ten; cruelty, bullying, lying, pain for pleasure, jealousy, brutality, murderous rage, molestation, deceit and torture. His authority was complete, nerveless, and surprisingly, compassionate. He was efficiency personified; judge and jury, and rarely ever wrong. He had already cleaned up one pocket of sadists on a nearby planetary system, but he didn't consider them the source as much as a branch whose root was somewhere else. He decided that this might be the source. Milo certainly displayed the characteristics for which Michael was looking.

There was a second part to Michael's commission that he especially liked. His second job was to create as many human/Aberian children as he could. He liked human women and they seemed pleased with him, so part of his job was very pleasant. As soon as he got rid of the source of the ugly then he could move on to the pleasant part. He chose women who he thought of as strong; who he believed would make responsible parents. Aberian children were born very strong so he was not really worried about them surviving, but he wanted a safeguard anyway. When he was sure the women he had chosen would conceive then he would move on. It didn't hurt any that he was immortal. He had a great deal of time to complete this part of his task, and he felt that he was leaving behind an ever growing group of beings who innately understood Aberian characteristics and ways. Then in a few thousand years, when the invasion force arrived, they could move right into the colony without bloodshed. Aberians had been given this quadrant of the galaxy to inhabit and they did not want to live where Timony had established his trash.

He didn't explain to anyone what he was or how he was different. Michael moved from place to place looking for groups of people who were breaking the accords and creating children whenever invited. He chose the disguise of Tom to be old for Milo's benefit, but he had given Wyette little glimpses of himself with a younger persona. They had developed a friendship over the last two months during the times that Milo had been away and Milo had been gone a great deal.

After Michael had taken care of the tractor, he carried Wyette to the house. He placed her on the bed while he prepared a warm bath for her. Slowly she came awake. The white plaster walls came into focus and she suddenly sat up startled, "Tom, you can't be in here until Milo is gone. He'll find us and kill us both!"

"Milo is definitely gone. Don't worry," soothed Tom as he slowly made himself look younger and more muscular.

"Did you kill him?" questioned Wyette quietly.

"No, I did not kill him. He is still alive, but he is sort of sleeping," responded Tom.

Wyette thought about that for a moment. "Will he ever be back?"

Tom shook his head no. "Not ever. I can show him to you if that will help."

She thought about it and then said, "Yes, Tom, I need to understand. Please show me what you have done with him."

Michael became younger still, and opened the pouch tied to his belt. His shirt came open as he pulled the pouch from his belt. He peered into the bag and pulled out three marbles dropping two of them back into the pouch. "This is him."

Wyette looked at the marble and back up at Tom. Her jaw went slack and she looked at him with large eyes. "Are you an Aberian police officer?"

"Yes, some call me that," responded Tom, waiting to see if she would start screaming.

"Are you here to catch the people that Milo spends every week end with?" she queried.

"I plan on taking Milo's persona and I'm going to join Del. Yes, I will capture all of them in Northfield this weekend. Do you hate me for what I have done?"

She exhaled in a sigh. "No, I don't hate you for doing what needs to be done. Please be careful and promise to come back."

It was all the encouragement he needed.

When he left her hours later, she was sound asleep. He was fairly certain that she had been impregnated. He had left her with a sense of security so she would not worry over the weekend.

As Michael went to his cabin, he could see large black clouds gathering on the horizon. A storm was coming. The wind turned up dust devils making it hard to breathe when moving from building to building. After he had his things, he went back to the barn. The electric carriage was not to his liking. It was slow and only carried a charge for so long. It had a short bed on the back which would help after he got back. Wait. Was he coming back? He did kind of indicate to Wyette that he was, but …no matter. He could make up his mind about that later. Then he focused on getting his disguise just right. Del was a lifelong friend of Milo's and he knew Milo really well. The disguise had to be perfect.

Once Michael reached the tracks, he could clip the old car into the power grid. He would stop at Del's to pick him up just as Milo would have, but he would feint a headache to avoid talking to Del or just put him to sleep if necessary.

The trip was uneventful and very quiet. Michael told Del that he had to go buy parts for the tractor. He said he would meet Del at the meeting place at the arranged time that evening. After Del left, Michael changed form again and followed Del to the meeting rooms. Michael observed the pass words and followed Del inside.

The building was two stories. The sleeping rooms were on the second story and the meeting rooms were on the ground floor. Michael learned that only members would be allowed in this evening. Two chapters were expected, the Ambassadors and the Black Widows and each had a different password. Michael decided that he had seen enough for right now. He would know this evening if he had come to the right place.

On his way back to the farm supply store, Michael noticed a group of people with their hands tied behind their backs being herded down the street. They looked confused and frightened. He wondered at the large number of prisoners. He asked at the farm supply, but no one seemed to know anything about it. Again as Milo, he purchased building materials to repair the barn at Wyette's steading as well as fencing and other supplies that he thought she would need. He made arrangements to have shipped all of the order which would not fit on the electric carriage. Then he found a clothing store and made arrangements to have baby things that he thought Wyette might need shipped to her place. He found himself looking forward to returning there. He smiled to himself as he realized that he had every intention of staying there for a while.

Later that evening, after he had checked in to his and Del's rooms, he found that these rooms were regularly reserved for them. Del excitedly chattered about all he wanted to experiment with as well as purchase. Milo changed his clothes and went down to the meeting rooms after he had placed Del's marble in his pouch.

There was an air of excitement as people set up their wares. The problem was the nature of the wares. One booth was filled with a wide variety of tools of torture. Another booth was selling drugs very openly and offering free samples of a new drug used for pleasure. Alcohol was also readily available as well as pornography of all kinds. Yes, he was sure this place and these people were the ones for whom he had been searching. He began to memorize faces and to categorize his plan. He studied every room and every booth. He left compulsive messages with owners of slaves to send them out in the town for one thing and another and he found the room where the prisoners he had seen earlier were being kept. Some had already been purchased for demonstrations. These he created an invisible bubble around in order to protect them. He also created bubbles for other people he considered innocent or coerced. All of his preparations had taken the better part of two hours. He walked purposefully toward the room he felt contained the source of the evil pulsing that permeated the whole building. The source was a large chest that seemed to emit a soft green glow and a musky sort of fragrance. The room was filled with people who were all naked and kneeling on the floor in a position of prayer or meditation. He quickly surrounded them in a bubble all their own with the evil box. Slowly, he reduced the size of the bubble until he felt the box recognize him. It shuddered and jolted as it diminished in size, but it was too late. Michael compressed this group quickly. He did not stop at the marble size, but kept going until with a pop all of the people and the black box simply disappeared. There were others that he eliminated completely, but most he compressed to marble size to be stored until he found a place in which to deposit them. Some would get a second chance, others would not. Michael created the illusion that the room and box were still there as he continued sorting, judging and compressing. Each compression was followed by an illusion until the place began to feel calm and empty. Michael snapped his fingers and the illusions fell away leaving an empty room with some rubbish in it.

He left the building and went outside to the area where he had put the bubbles of the innocents. He released them. Then he told them that they were to tell the rest of the town what had happened here. He decided it was time that this planet know an enforcer of the Compromise had been here and passed judgement. That should discourage any he had missed from following this path again.

As Michael returned to Wyette's steading, he changed back into a young version of Tom. Looking forward to good honest hard work and a pleasant relationship gave him a sense of satisfaction. He was not a braggart, but cleaning out this nest had felt like a really big step in the right direction. He was quite sure that none of the people he had passed judgement on would be missed. Cruel people were feared, not loved. The planet now stood a chance of growing and prospering in a healthy environment, and he would, soon enough, be a father. Maybe he would stay long enough to come to know this child.

Relationships

Environment

Your personal psychology is impacted by your environment. Rarely does abuse ever effect just one person. If one person in a family is abused then the other members of the family either feel guilt, arousal, shock, fear, or any of a number of other emotions. All of this emotion creates an environment. Environments that are filled with negative emotions are not conducive to healthy human growth. As humans we need to feel loved so we can grow. Our growth doesn't stop when we graduate from adolescence. We should be growing and becoming all of our lives. If we aren't, then something, usually abuse of some kind, happened to us which stopped or slowed the momentum of our forward progress. Those are the obstacles that we need to overcome in order to be happy and to continue our growth.

Sexual abuse is the first kind that I want to address here. My grandfather became an abuser. I don't know the real reason behind this happening, but I have thought of a group of possibilities that might have caused it. His father died when he was a child. He had two sisters and his mother was very hard pressed to care for all of them. He had to go to work very young to help the family. This may have been too much responsibility too young. He developed an inferiority complex. Maybe his exhausted mother had a sharp tongue, or maybe he was sexually abused himself. Also, society set up a way for him to get by with sexual abuse. It's called the paternal system. The father of the house is the God figure of the home. He is the absolute ruler of his domain. All pay respect to him as he is the sole provider for the family and he always knows what is best for everyone living under his roof. In the case of my grandfather, the women of the house covered for his sins because they believed it was their duty to do so. My grandmother was uneducated and could not support four children on her own. Besides that, my grandmother was brainwashed into believing that my grandfather was God's representative on Earth. By definition, he could do no wrong. She loved him unconditionally. Nothing, not even his abuse of her family members shook her confidence in him. That might have partly been caused by things he said to her like: "Women have to learn their place. They have to cooperate. They might as well learn young." Or "You know very good and well that it doesn't hurt them." or "I never hurt my girls!" and "Woman can't do anything else, so they ought to be good at sex." (In case you are not sure, all of these are lies.) All of this created an environment that made sexual abuse possible. I think it likely that he was doing just what he wanted to do and was blaming being drunk as an excuse.

I believe that my mother repressed his sexual abuse of her. She chose to move our little family two hundred and fifty miles away rather than confront him with the abuse issue. My aunt would get a sad look on her face and just say, "We have to forgive and not carry grudges." Her marriage ended in divorce. My mother's relationship with my father was extremely dysfunctional and I suffered through four divorces of my own. My cousin also has had problems with relationships. I think that is a lot of collateral damage for the urges of one man.

We have to first admit that there is a problem before we can fix a broken environment. If you suspect that something might have been wrong with your childhood, I urge you to get professional help in uncovering any issues that you might be suppressing. These issues need to be faced if they exist, but it is very difficult to do it alone. If your current situation is abusive, get out. Get help. If you are truly searching for a way out, you will find a path to take. It is important for you to get help to find peace of mind and a new format for life. Abuse is not the norm and having a healthy environment to live and grow in is absolutely necessary. Abuse is contagious and not something you can put on a back burner and let simmer while you tread water and do six other things. It comes back to haunt you. It shows up in your life in the most unexpected places. Please, do not tolerate this.

The other area of abuse that I consider extremely damaging to human progress and growth is lying. There are many kinds of lies that permeate our society. We are the species of the lie. There are religious lies, white lies, lies that we tell ourselves, and lies that we tell to protect someone else.

Religious lies come in all sizes. Mary Magdalen was not a whore, but the old Church lied that she was to keep her ideas from being respected. Mary Magdalen supported Christ and was the closest of his disciples. She was influential in early Christianity. How about the lie that men speak for God, or that forgetting is part of forgiving.

White lies are the ones we tell our children because the truth might be too much for them to understand, or the lies we tell our best friends when we are trying to avoid hurting them. We also lie about the Easter Bunny, Santa Claus, and the tooth fairy, but we do that to create magic in our children's lives. I'm not saying we should stop loving Santa Claus. I'm saying we should just be honest about it. We are pretending. He is an idea we love. He lives in our hearts and we all pull together to enjoy this idea. It is a good thing to pretend and make believe kind and generous things. Therefore, Santa and his elves, gifts and giving are all or part of the season that we make as joyous as possible. It doesn't take anything away from the religious events of December to be pretending that goodness is abundant. Honesty includes instead of separating. Honesty highlights importance and allows for children young and old to enjoy the festivities.

Some lies we tell ourselves: "I'm not really that bad looking at this weight.", "I'll stop smoking as soon as ...." Or "I don't really need to wear a helmet when I ride my motorcycle."

Other lies are to protect others. Most of these, even the well-intended, can end very badly. Once a person figures out that you've lied to them for any reason, trust is negatively impacted.

Please, do not lie. The truth may be hard to face, but it is always better to face reality and make the best of it, rather than dealing with the consequences of lies for years and years. You cannot grow if you are lying to yourself or others. You must make this change. We base decisions in life on known information. If the information we know is from lies, then the outcome of our endeavor will be false as well.

In my own experience, my family was ripped in two when my uncle caught my grandfather with his hand inside my underwear. A man is harder to bamboozle with lies and rhetoric than lowly women were. Still, the problem was not addressed, just ignored and my uncle took his family away to protect them rather than address the lies that had stood unchallenged my whole life.

There are other very damaging behaviors that we often overlook as socially acceptable, and they should not be tolerated. Teasing (whether friendly or not) should be avoided. Bullying is obvious, but there is also gossiping which can be just as destructive. A person with a sharp tongue can cause damage to another that might never heal. Do you really want to be around people who look for fault in others to puff up their own self-image? These are all damaging to an environment and they appear to spread like disease from perpetrator to victim. When people are abused and nothing happens that stops the wrong or even explains that it is wrong, the victim comes to believe that the behavior is justified. They may even become abusers themselves. People who are constantly embroiled in drama are also troublemakers. They broker worry and stress. If I have just described the people you hang with, then you need to look for a different crowd. You need to find a group that demonstrates the following behaviors:

1. Calm, level-headed thinking when problems arise.
2. People who are supportive family members.
3. People who live what they believe.
4. People who meditate know how to address stress and angst.
5. People who are emotionally centered.
6. People who speak positively instead of looking for fault.
7. People who are respectful of others and their property.
8. Creative people.
9. People who have fun (laugh a lot) without making fun of others.
10. People who are generous and inclusive.

You will find that these people have a joy for living that is contagious. It is hard to be sad or brooding around someone who is full of joy and enthusiasm. Avoid the people who are not making any effort to improve their emotional environments.

Chapter 6

Environment

STORY

The gray chilly station bustled with people coming and going. The immortals were very busy. Each had assignments and most required travel. They traveled light with a kit provided for them. The kit contained nutrition in pill or powder form and a small wrist unit that produced pure water from elements present in the air. Each carried at least one AIU (Artificial Intelligence Unit). These units responded to voice command and became whatever the possessor requested of it. For instance, the owner might need an emergency space suit or they might need a comfortable place to sleep or a conveyance of some sort. The AIU's were very adaptable and surprisingly dependable. It was almost unheard of for one to malfunction.

Michael was going to a hostile environment so he was taking a full ten AIU's with him. He liked his comforts. He was in the process of checking his pack when Timony approached.

"Greetings, Michael," cooed Timony. "Where are you headed?'

Michael nodded acknowledgement to Timony. "I'm going to Pirida. There have been reports that the Urki are trying to expand into that galaxy. How about you?"

Timony straightened his back and did his best to sound casual as he announced, "I'm headed to Earth. I'm to be assistant to the Controller there."

"That's an influential post! You do know that the current Controller is Arkian, don't you?"

"Yes, of course. I'm looking forward to the challenge. You know me; always looking for a fight. Hey, are you and Athael still seeing each other?" queried Timony?

Michael tried to hide his surprise as he turned toward Timony to answer. "No, we just haven't had downtime together and so we've grown away from each other. Why? Are you interested in her?"

"Well, we grew up on Raline 4 and we've sort of kept in touch over time. I left a surprise for her at Central. I was just wondering if she'd gotten it yet."

"No idea, Timony. Well, I've got to meet my connection. See ya around."

Michael remembered Athael telling him about Timony's mean obsession with her. He liked to play tricks that he thought were funny and lately his tricks had become more mean-spirited. Michael feared that Timony was on a destructive course and might try to hurt Athael. The decision was quickly made to pass by Central before he made his connection.

Michael, a shape shifter, had the ability to change his form to that of any living entity. He could not become a rock or a rocket engine, but he could become anything from an earthworm to a flying dragon if he so wanted. It was a characteristic of those born on his home planet of Abare. He loved his AIUs. He thought of them as cousins. Grabbing his favorite AIU, he rounded the corner of the station. As he willed it, a sleek black racer appeared and he became Athael, in black leather, riding it.

Central was as crowded as the station had been but Michael had no problem imitating Athael and obtaining the sequa sealed box containing Timony's present. He left the station by the nearest door and assumed his natural male form as he walked. Breaking into a run he spotted an unoccupied bench in the center of the adjacent park. The box must not hold explosives as something solid moved from side to side as Michael ran. He easily broke the seals and opened the container. He found Tipcan, Leona's pet in a comatose state within. Michael quickly drew in energy from the park around him and gently picked up and scanned the still form of the rappa. He was alive, but just barely. Slowly the shape shifter infused the small animal with life essence and watched as it became stronger. Soon

Tipcan wiggled and squirmed to get down and run. Michael played with him there in the park until Athael arrived about an hour later.

Athael was surprised to find them together. "I came over to pick up a package. "Wait, I thought you were leaving today?" questioned Athael puzzled.

Michael chose honesty as he told her that her package had been from Timony. Her eyes became large with shock as she put it together. "You saved Tipcan for me, but how did you know?"

"I ran into Timony at the station. He mentioned he had left you a present. I assume he was going for a "two-fer" by planting the information and then hoping for guilt when I heard of the death of the animal. Tipcan would not have made it until you arrived just now. That man has a sinister mean streak, I'm afraid. Whatever did you do to him to make him hate you so?"

Athael took a deep breath and sighed. "I know it's hard to believe, but we were best of friends when we were children. We spent a lot of time together until one day the sky turned dark yellow. There was a sulfurous smell in the air and Timony said that he needed to get rid of his anger. He said he was going to tear up Widow Martin's garden. He wanted me to join him. He said it felt good to him to be a destroyer. It made him feel free to destroy whatever he wanted. I was shocked. He didn't sound like himself and he frightened me. Upset, I ran home. I was crying and shaking as I burst into our kitchen. Mama was washing dishes and I could smell soup cooking on the stove. She asked me what was wrong. I sobbed out the story. We went to his parent's home and she insisted that I repeat the whole thing. Then we all went to the widow's house where Timony was still in the heat of battle with the cabbages. He never forgave me for going for help."

Michael shook his head, "So he's gone from bad to worse."

"He's made one bad decision after another" bemoaned Athael. "He thinks it makes him stronger, more powerful to destroy and torment. He avoids creation and he always has his eye out for the barb or word that will cut or twist a situation.

"Michael rubbed his chin, "He's headed for Earth now. So maybe you've seen the last of him.

Environment

Faith, Religion and Knowledge

Faith is belief in something unseen, unproven. We are happier and stronger when we have faith. People who have faith live longer, and many would argue, have more satisfying lives. "Faith in what?" you might ask? The Bible argues against itself. More than half of the New Testament has been proven through modern scholarship to have been forged. Early printers changed text on accident or on purpose. They kept in parts that they knew to have been tampered with, but left them in because they wanted to be the first to publish. The freshest versions we have of the New Testament were written down for the very first time about two hundred years after the events took place. And, they were written by the poorly educated; women and slaves. Also, monks copying the letters added and deleted until we just really don't know what is original and what is not. It would be so neat and tidy if the whole Bible text could be accepted as truth as written. However, legend, storytelling, embellishment or out and out lies have all been mixed together with truth. It is very difficult to sort out. This is not to say that the Bible is not of value to people in general or people of faith. Quite to the contrary, the Bible, quite miraculously, remains a source of inspiration and contemplation. It has historical reference points and many of its ideas improve life for millions. I have been searching for answers to needs in my own life and opened the Bible randomly. Doing this I have found answers to my queries or comfort in my immediate need. Positive spirit can achieve wonderful results. Have faith, but don't judge others based on the printed word. You need to have the knowledge of the Bible's weaknesses to put into perspective using it. Personal use, great; condemnation of others, not so valid.

We may not know exactly what happened back in the days of Jesus of Nazareth, but we do know that something enormous, precedent breaking, did happen. For one thing, people started talking about forgiveness for the first time and something happened that got people talking and responding with faith. I'm not pushing theology here. I'm saying that whatever happened back then was major because we are still feeling the aftershocks of it. Whatever those events were in detail we cannot know. What we know is that it changed the world. Many religions have been formed which claim to know for sure what transpired, but really they are guessing, making it up, or have faith that what they preach is truth.

Dogma and rhetoric may have to change or morph some, but I'm pretty sure that organized religion is not going away any time soon. The more knowledge we obtain the less workable highly structured religions seem until you consider how necessary houses of faith are. We need these organizations as centers of community activity. We need places in which to gather and share spiritual awareness and study. We need places to celebrate life's experiences and to organize help for those in need. However, we also need to realize that these constructs can take on negative attributes if we let them. Many current religions are male dominated and therefore, lean toward aggression. This is a warning not a condemnation. It is not my issue here to place blame, but to give people a "heads up" when associating with organizations that might use the Bible as a reference to hurt an individual or groups of people. If you are a person who believes in all things positive, then don't join an organization that preaches love and practices hate.

I applaud the religious organizations who encourage and demonstrate balance between male and female energies. People who run the soup kitchens and food pantries are deserving of praise. Those who raise money for needy families and help those who have suffered from natural disasters deserve praise. This is as it should be.

Archeology has been and will continue growing, expanding and stretching as a discipline. New equipment, better training, more interest in the subject and a constantly growing base of knowledge provide a fertile environment for growth. The movement called Ancient Alien Theorists have amassed a very large bank of interesting facts that point toward the existence of alien intelligence, greater than our own, influencing (you might say growing), guiding and uplifting mankind. All of the ancient religions reference a time when "Gods" walked on Earth with man and this is from every corner of the globe. We have found signs of advanced civilization as well as signs of nuclear warfare, which are thousands of years earlier than cavemen. As this new scientific information comes in, we will need to adapt our perceptions of both what is alien and what is religion.

Most modern Christian religions still teach that Christ will return. He arose from the dead, ascended into Heaven and traveled somehow to his home where He is preparing a place for his chosen. Doesn't that make Him an alien by definition? He was half human and went to live in the stars for a few millennia. When he returns, it is his plan to take a number of humans off world for a three-year party and wedding celebration while the Earth goes through a destruction of some kind. This sounds remarkably like an invasion with purpose. Then there is to be a one thousand year reign of peace while Satan is bound. Mankind is to be given the opportunity to adapt to alien leadership and to decide if they want to live in a world without evil influence. It seems a bonus here that the aliens have the technology to eliminate death. They proved this by resurrecting Christ. Since they can resurrect the dead, they will probably choose who to resurrect pretty carefully.

Even though the aliens are planning on letting the planet endure a destruction, they seem to believe in love, peace, grace, goodness, and forgiveness. They have given thousands of years of warning and laid out their entire battle plan. This speaks to incredible confidence. This destruction appears to be necessary to rid the Universe of violence, hatred and aggression of all kinds. What catches my eye here is that the number of humans who choose to remain on Earth will be as numerous as the grains of sand by the sea. Personally, I'm on the side in favor of goodness in this scenario. I don't care for violent, mean spirited people. "Yay rah, team alien!

All of this boils down to the most basic of human emotions: love versus hate. An act of violence which is perpetrated against us functions like a wave of extreme negativity. Think of it as if we all lived in plasma. The wave of violence hits us, but hitting us does not stop it. It impacts us and others around us each time we remember the incident like the waves of a tsunami. Like a gift that just keeps on giving. In order to recover we have to stop the wave motion. Talking to a counselor, journaling or just plain writing about it, not just once, but over and over again. If we want to be happy ourselves we have to step away from violence. We have to separate ourselves from the cycle. We must become bigger than the problem.

One idea that can help in this process is realizing that people who perpetrate violence are not happy people. They are sick people who have been abused themselves and have just agreed that continuing the cycle of abuse is what they want to do. They choose it. Seeing them in this light makes it easier to back away emotionally from them. You are not responsible for changing their thinking. You are only responsible for your own responses, actions and deeds.

I'm one of the people who believe that we live in a shared Universe. I believe that one self-aware, omnipotent force (which I call God) created all that we know and more out of itself. This force is in us and all around us. It creates, energizes and empowers.

We live in a petri dish of sorts. We are an experiment which has almost reached its conclusion. What have we grown into? If cruelty and selfish pursuits are our bread and butter then we will be found out by the ripples of the waves we have left behind us. If you move to a new neighborhood do you look for dishonest, mean-spirited, vengeful, unhappy people to live near? I do not believe that our neighbors in Orion's Belt or Cygnus or Sirius will be any more likely to choose that sort of neighbor than we would. The major difference here is that they have the ability to eliminate the lousy neighbors, and will probably relocate those who have not yet decided which church to worship in to another negative planet until they can decide.

To summarize: faith and religion will need to grow and change as knowledge brings us new information. What are we capable of accepting and what is our bottom line in matters of faith? Do you believe we are not alone in the Universe, or are we worshipping in the money cathedals?

Faith, Religion and Knowledge

Terra awoke groggy from the anesthetic. Her mouth and nose were dry from the oxygen tube. Slowly the sterile pastels of the walls and curtain began to come into focus. Her old fear returned. What had they done to her this time? She took attendance of her body parts. There was pain between her legs. Her nether regions both ached. Her legs felt weak and hard to move. She could not feel the incision that the doctor had made for the back surgery. At least *his* work was not an area of concern. She was thankful that she could see. As she cataloged her aches and pains, she remembered the time before her last back surgery when the crazy nurse had pulled her from the gurney after her myelogram. The crazy nurse was fiftyish and wore an old style military nurse's uniform. She wore a small gold cross at her neck. Her permed dark brown hair was cut short, styled in tight curls to her head. She had grabbed Terra by the front of her hospital gown and said that they had to get out of there, and began pulling her down the hall. The good nurse that had been assigned to her had been called away for some unknown reason. The crazy nurse got her to sit up before the good nurse could stop her. The screaming began right after the crazy one pushed her back down and said, "That should do it." She smiled a hate filled smile and turned and ran. The good nurses did all they could to help Terra, but the intense pain did not give up easily. Her eyes had been light sensitive ever since. The eye doctor had called it glaucoma.

She was pretty sure that she would need a D & C and colonoscopy to correct whatever they had done to her this time. Her mind continued to wander back to other times of attack. There was the time they put snot in her coffee at the gym. That one had caused a serious sinus infection and infections in both ears. It had taken six months to recover from it. Then there was the time they had sprayed a surfactant on the shower room floor at the gym. She had slipped and fallen hard breaking her left wrist. Oh yea, there was also the time at the art museum when a woman very similar to the crazy nurse had blown germs of some kind into her eyes. It had only been a month or so to recover from that one. The guy that jacked off in the swimming pool while he watched her swim laps was, at least, not damaging to her physically. Then there were dead taxidermy song birds left on her front porch and on the walkway into her place of business and the time they blew up the tire on her truck and two other times that they ruined tires by screwing large screws into them. Who could forget the constant damage to the body of her car, and the poison ivy juice in her face cream, the razor blade cuts on her heels and being followed everywhere she went. There was the time they had used an air gun at a restaurant to shoot her in the heel with some kind of plastic bullet, and the time that they had set off a germ bomb at University Park leaving her to sit in the germs waiting for someone to take the cash box. There had been so many attacks that she had given up recalling all of them. After all, it didn't really matter. She always healed and was not depleted but strengthened by these bazar attacks.

Timony had been hard at work making her life on Earth as difficult as possible. Her consolation came from knowing where Range was. He was on the right track, and he had recognized her. As long as he could do what he came for, that was all that mattered. She would endure the perversions of the sick ones. Somehow, she always avoided the worst case scenarios and she always healed quickly. Range would know that she loved him and that a sacrifice had been made for him.

In another part of the galaxy Michael was sitting with Terra's Higher Self.

"Michael, please consider this contract seriously. I have agreed to send my my avitar to Earth to help with Range's mission. With Timony there, I am going to need protecting. Will you please accept this responsibility?"

"If I accept your contract, Atheal, I will need to send an avitar as time moves differently there. I will need to stabilize the continuum as I have been required to show up there for the coming Harvest."

Higher Self Terra was quiet for a moment, "Earth is moving into fourth density and Timony will not do well in that vibration. He will be desperate to keep the numbers low for Harvest. My aviator will fight but she will not understand why she has to face so many negative entities. I have been sent to a deeply religious family and you know what a tangle negatively polarized religion can make. Your avitar will need to go back in time to her birth but stay in the fifth dimension. You should be able to monitor her life and impact if necessary. Timony will be leaving about the time you are arriving in real time. I think this can be a workable plan. Will you accept?"

"I will accept the contract, Atheal, as I understand."

"And so it shall be."

Destiny

Have you ever known,
Somewhere in your heart,
That you were destined to do something?
Could feel it deep inside?
The drum of your soul beating out:
Learn……. process, Learn… remember, Learn... become, Learn…
grow.
Each lesson a step closer.
Companion impatience natters hurry, hurry.
You must keep going. It is your fate.
Prayer has become mindless begging.
You have worn your smiles thin.
You have cried until tears were only rainfall.
Bent by the weight of what you carry.
Turbulent anger teakettles with anguished cries for revenge.
Do the heavens see?
Do they know of this malice?
Does God care?
Maybe.
Rise above tragedy, injustice, loss and pain.
We come to know order better by what it is not.
Death has no meaning, or suffering, or pain,
Or space, time, matter, or loss.
All are meaningless but love.
Love, vague as it is, continues unfettered, knowing no boundaries
and having no license, yet building tabernacles of continuance.
Forgiveness buys the ticket.
Is creation served by what must be left behind?
Or do we move forward on these steps?
Choices and their consequences being the true stuff of our lives.
Knowing this….What good? What destiny?
Oh, keep moving, one step at a time.
It is promised. The sun will shine.
Love will win.

Rise Up You Phoenix

Are you brought low?
Has your soul burned up from self-sacrifice?
Have you been wrong; wronged; abandoned?
Have you gone down the wrong path looking for pleasure, for love, for self-esteem, for glory or for eternity only to find your way blocked by mourners, critics, those who laugh at your efforts, or those who hate you for trying?
Is there a deep, low, mocking voice that laughs at your pain filled misery and takes gratification from your suffering?
You have made bad decisions that have caused others to lose confidence in you. You have hurt people who love you and you have not been capable of being everything to everyone. You have thought that you do not deserve to be forgiven.

    Rise up you phoenix! Rise up I tell you!
Uncurl your punished body. Slowly,
Peacefully open yourself. Open your eyes to the light.
Accept your red swollen wounds.
Unwrap the throbbing aching mass that is your heart.
Extend your lonely being and feel that there is good.
Feel the top of your being open.
In your mind see light and let it pour over you like warm rain.
See yourself rising up in this primordial rain.
Rise up you phoenix and bear your burden again.
You can.
Yes, the light after the rain will be nourishment.
Drink it in and the light will satisfy.
The light will bring you rebirth, a pattern repeated; a tesseract.
You can do it right this time.
Forget the past for a moment and look where you are going.
You know what it is you must do.
The suffering is only as deep as you allow it to be.
Teach by example, entertain us, lead.
Put one foot in front of the other.
When you feel your momentum take you high,
Look back for an instant and admire how far you have come.
Sail, soar and glide in the sun.
What you have done is to forgive yourself. That was the launch.
Rise up you phoenix.

Continuance

Smile.
Stretch.
Move.
Keep walking.
Organize.
Heal.
Forgive.
Grow strong.
Become.
Glow.
Shine.
Give back.

Faith, Religion and Knowledge

Sexuality

Four Rules of Sexual Encounter

1.   Be old enough to take care of yourself.

That means: Provide your own food, insurance, shelter, clothing and transportation. If you don't follow this rule, the result is poverty, depression and sometimes crime.

Why would you take a chance on bringing a child into a life of deprivation? There is one other reason for this rule. The cycle of sexual abuse of children has to stop. Young children are very sensitive sexually. They have the ability to climax almost indefinitely. If they become sexually aware too early, they can spend the rest of their lives trying to duplicate early experience. Predators take advantage of this. How many people do you know who have not been able to maintain a monogamous relationship? Ask them how old they were when they had their first sexual experience.

2.   Always be kind and loving.

Never, never, never have sexual intercourse when you are angry with your partner. Make peace first. If you can't make peace, abstain. Anger opens up the door to something else and sexuality becomes a destructive tool.

3.   Experiment.

Be sensitive to your partner. Experiment slowly. Learn to build sexual rapport. Take the time to tantalize, and you both will reap benefits. Grow in your sexual life like you do in your relationship.

4. Don't cheat!

It is rare that a relationship can recover from betrayal. Once trust is broken, it is almost impossible to reestablish it again. The quality of what you once had will never be there again. If you value your relationship, be faithful to it. If you don't, then end it. It is not possible to experiment with multiple partners and build paradise with one person at the same time.

Humans require love to be happy. From birth to death, we will fail to thrive without it. Love and sexuality are not the same thing. Love is the giving and receiving of good and it is always kind, generous and compassionate. Sexuality, on the other hand, can go either way toward light or dark. Sexuality can be the most wonderful of shared experiences or a complete nightmare. Our society is so fearful of the power of sex that they lock it away behind doors of secrecy and ignorance. Worse yet, society has let advertising (with obviously ulterior motives) take over the sexual standard by benchmarking what is beautiful. They, not us, decide what is attractive and sensual. Models, rock stars, movie stars and wealthy partiers are the examples we let stand for our young adults. I'm not saying that all of these examples are always wrong or evil in some way. I'm saying that as a society we should be more honest and open about these things. Young people should be taught that a truly beautiful person is much more than an attractive exterior. Beauty extends to having a great sense of humor, beatitudes, achievement, creativity, ability to communicate well, an ability to impart joy, and so many more qualities that a person can possess. We allow violence to fill our children's lives in video games, television shows and at theaters, but healthy sexual relationships are never demonstrated. Is violence really more acceptable than healthy sexual relationships?

Perhaps there is a problem with understanding what is healthy or normal? After all, the end goal of sexuality is to procreate. Depending on your religion, there are sometimes strict rules concerning procreation, but saying that and just leaving it to religion to set standards for sexuality based on procreation is like letting a small child run the house.

We could start with doctors providing a variety of information to parents at their child's tenth birthday. It could actually become a health based initiative. We do, after all, live in an information society. Surely there is someone out there willing to prepare and print information on the topic for a variety of different groups. It's not a complete solution, but it would at least offer a beginning for more open and honest discussion. We could also discuss the Rules of Sexual Encounter that I began with in this chapter. Give our young people better examples than crass, out of control, young adults. Or worse yet, the interference of predators.

Chapter 8

Sexuality

Terra shook out her long brown hair in preparation to dry it. Her heart was still racing. The water had been cold and had felt so good. Cool, crisp and supporting; it balanced her as she crawled through the water. Today had been a shock. One of the other swimmers, who was swimming in the lane next to hers, had been rubbing his penis as he swam. She was shocked that he would do something like that in a public place. He was a full grown man, not a fourteen year old and she knew that he had a wife who also sometimes swam in the morning. Come to think of it, his wife was pregnant. The manager of the pool had to ask everyone to stop swimming and get out of the pool so he could clean up the mess this guy had made. The whole thing had left her feeling confused and used somehow.

It reminded her of the time she had gone to her family doctor for a pap smear and the doctor had used long wooden cotton swabs to climax her after the procedure. The climax, her first of that kind, had felt good, but again it had left her feeling used. He had not even asked permission to touch her in that way.

There was another time when one of her pastors had asked her to join his wife and him in bed. Well, at least he asked.

Another time a close friend had started masturbating while they had been watching TV. Lucy had been at their house for two days while some work was being done at her home. Another time, the husband of a dear friend had asked her to have an affair with him. She kindly declined the offer. Was it like this for all women?

Men were so willing to make and break rules. Did that mean that the rules weren't really that important or that there was some weakness in all the people she knew? Being in her forties and single had not left Lucy with a lot of social opportunities. She actually understood why her girlfriends did not want to include her in events. Their men were all out of control. But where did that leave her?

She felt very much alone. So far, her choices had all been cheat or do nothing. She had been choosing nothing, but that was getting old. She wasn't getting any younger and there was really nowhere to turn. The only conclusion she could come to was that somehow she represented sexuality to all of these people. Perhaps she would move and make new friends. It was at least worth considering.

She tipped her head sideways to try and get the last of the water out of her ears. She could still smell chlorine and her fingertips were wrinkled like dried prunes.

As she approached her car on the way to work she noticed a dead canary lying in front of her car door. It made her feel sad and then she noticed that the bird wasn't just dead. It didn't have eyes or guts or any of the parts of a bird that had recently been living. As a matter of fact, it had been preserved and stuffed. Wow, now that is powerful strange. She pondered the mystery all the way to work.

After work, she went to the grocery store to pick up a few things. While in the grocery store a woman with an ugly expression on her face ran a grocery cart into the back of Terra's heels as hard as she could. Terra looked at the woman running away and wondered what she could have done that would make another human being act that way. The heel of her foot was bruised and it caused her to limp. Terra decided that the woman must have been deranged to attack a perfect stranger like that.

The next day, on her way to work, she noticed a car parked alongside the road. The following day the same thing happened only in a different place. Every time she went into a store to shop, the shop became crowded before she left. This was also true of restaurants, doctor's offices, the library and public events.

By the time a year had passed, Terra was sure, beyond a doubt, that it could not be coincidence. She was being watched in great detail, by many people. What she could not figure out was why anyone would want to watch her. Perhaps she could ask them? She tried, but they seemed to move away from her if she approached, or became involved with some other person, or just vanished when she turned away for an instant. She even considered that she might be losing her mind.

These weird occurrences became Terra's normal. The ebb and flow often changed. At times the flow would only be a trickle of strange, and other times it became a raging torrent. Why didn't they lose interest in her? She was so boring. She did such everyday things and, was not particularly attractive. She didn't have anything extravagant going on in her life. She didn't date or have a love life of any kind. She was not illegal, immoral, or a criminal. She went to work, paid her bills, and lived quietly. What was going on? One conclusion she had come to was that whoever was having her watched was not particularly concerned about her well-being.

One morning, Terra awoke from a dream that had frightened her as it had seemed so real. In the dream, a consciousness had told her that she would tell the world about things that were going to happen; a destruction was coming. The Earth would be destroyed by fire. All that was third density would burn in this fire. An attempt would be made to save all of the human population of the Earth. The people of the planet would be sorted into groups and then taken to different places to live. To be saved from this destruction and go to a positive place was to wear forgiveness like clothing. To be saved, was to accept a new order of compassion and love. Individuals must take on a loving, kind, peaceful demeanor and rise up out of the rage, hatred and chaos. Those who accept will be reborn into new bodies that are free from illness, deformity and death. They will enter the cosmos as new beings, retaining past memories of joy and of trial. After all, it was the worst circumstances that gave the best opportunities for greatness.

She vowed to herself that she would find a way to tell her story and this message.

Sexuality

Achieve Something Memorable

I've heard it all my life that I should do something memorable for mankind during my lifetime. Memorable is different for every person. What is memorable to one is boring or repetitive to another. Bottom line is that I need to achieve something that I believe to be memorable. This isn't for anyone else's happiness directly. Firstly, it is for my own peace of mind. Secondly, it is hoped, that what is accomplish will be of benefit to many. If we all took this seriously, as a society, just imagine the great good that we would accomplish.

Now what does achieving something memorable look like: I've considered using my voice to touch the hearts of the world. Or perhaps building something spectacular could fulfill this commandment. Planting majestic gardens to sooth the hearts of a weary world would do the trick. These are all large and lovely ideas, but not everyone can do large and impressive things such as these. I've decided that achieving something memorable might, more likely, mean devoting precious time to a worthy cause like a food pantry, homeless shelter or a soup kitchen. Or, maybe it could mean raising mentally healthy, socially well-adjusted children to improve the stability of our society. Perhaps you are a person who can create a foundation to help Latin American children further their education, or a person who can motivate millions with beautiful rhetoric. I once read a story about a lady who rode her bicycle all over Maine planting lupines in every wild area to make the world a more beautiful place. I can see wonderful merit to all of these ideas and in a multitude more of good and worthy projects. Pay it forward as often as you can. If we each contribute to good in our own best way, this planet will become more compassionate and less brutal.

I believe that it is important for you and I to show the world what life can look like when compassion and intelligence combine into action. Posturing needs to be left behind and clear eyed focus applied to serious flaws in our society. Our minds must be open to change and we need to have the end result in mind instead of partisan politics. Who is angry with whom, no matter the reason, is not important in the scheme of things. What is important is that progress is made which improves life's quality for all people.

For example, if we put a psychologist in every elementary school and provided before and after school child care in every building then local people would be in charge of the atmosphere within their schools and long range, within their communities. Parents would get the help they needed parenting and it would give them support in times of crisis. In my estimation, this would be a better use of tax payer dollars than building a bigger bomb. Many people have decided to home school because they have lost faith in the industrial model of education. The unfortunate part of this is that we might be losing our schools as centers of community resource and as a forum for learning healthy social interaction. If schools would focus more on the mental health of their students and their families, we would have less violence and better cooperation in the community. Currently, the schools seem to focus on numbers and scores and as a result have less time for the wellbeing of the actual students. Tax benefits could be offered to companies who give time to employees with children to volunteer in their schools. This time could be banked so that some people who wanted to use these days for other community projects or volunteering would have the opportunity to offer their talents to the community. I would suggest two days a month be made available for this kind of community involvement. These are just examples in one small area of American life. Just imagine what we could accomplish if we all were looking for positive ways to improve life quality instead of living in fear and anger.

Achieve Something Memorable

Eternity is a long time. To have an eternal body would be a source of great joy and power. Those who have this gift should carry it well. If all diseases are curable and the body never wears out, then an entity could continue into time and space leaving a very long trail of either good or evil. What would you do with eternity?

Michael went on to Piride where he wiped out the invasion forces who had been bullying the Piridians. He had sired many children and enjoyed a woman's company for all of her long life and that of their children for three of their generations.

Athael stayed at the mountain and finished her training. When her training was complete she stayed and helped to train young immortals who had come to the mountain for special training. It was rewarding work and she touched so many lives with her contributions.

When Michael came back to the mountain, he and Athael applied for assignment together. They went to the Abarian System where they helped to strengthen the central government and they created a lovely life for themselves while they raised a child into adult years. It was a happy and productive time for them both. Their son, Mika, chose to serve in the military. He passed before his twenty seventh birthday. He was an adult when he passed so he would not be coming back to them. He would choose his own next assignment as an immortal. After his death, Michael and Athael decided that they needed a change and time to adjust to their loss. When they got back to Sirius, Athael went to work as a grower of souls in the Eternal Garden of the most High. It was a dream assignment. Michael was very happy for her. He received his own assignment of traveling the Cantalorian System looking for seedings by the Arkians. They were both pleased with these assignments and set to work with vigor.

Timony, on the other hand, was leaving a different debris trail. He had angered the Arkian Central Manager at every turn. The Arkian, named Tole, loved to belittle Timony when he could catch the budding demon in what he considered to be an error. Timony was learning new and unusual ways to hate. He eventually figured out a way to eliminate the domineering Tole. Timony then became the new Central Administrator of the Earth Station. His pride and his hatred knew no boundaries. He knew that at some point, Athael and Range would come into his domain. Killing Range would be his joy of several lifetimes and he would torture Athael in every way possible. Timony felt sure that Michael could not protect Athael once she was on Earth. Timony feared Michael and would avoid confronting him if at all possible. The one he most wanted to destroy was Range. Range was Athael's pride and joy and losing him would cause her the sweetest anguish. The immortal demon set the pneumonic canon on a target setting that would fire as soon as Range entered Earth's atmosphere. He would be annihilated, not just killed. There would be no rebirth for him. Range would be a threat no more and his expiration would vex Athael immensely.

Athael entered Earth's atmosphere and saw the cannon open and come up out of the earth. She felt panic that she would be too late but she did not hesitate to draw all available energy to her and speed toward Range's craft. He would never know what happened but she would divert that blast and save him if she could. From that irretrievable moment she knew no more until she came awake into a very pink small space that she just knew she had to get out of in a hurry. It was an inauspicious entry into a cold life on Earth.

Achieve Something Memorable

Prayer and Meditation

I read a scientific magazine article that said that physicists have observed that we seem to be linked to our universe in a creative way. Astronomers study a section of the night sky looking for new stars and they seem to appear where there were none before. Those that study life forms continue to log new and unusual new forms all the time. People of faith believe in prayers being answered. We seem to be able to achieve what we can visualize. I believe that meditation and prayer are very similar and I think intense longing can be added as similar as well. Focusing the intensity of our being on a subject brings that visualization closer to reality. Some people can so control their mind that they can achieve a state of consciousness that opens them to genius, euphoria, or freedom from their own body. We are at our best when we are harnessing these abilities. There is so much we need to learn yet about the human brain and its capabilities. Or, maybe it is consciousness that we need to become more aware of using.

In my youth, I spent time in prayer. I prayed when I got up in the morning, before I walked out the door to go to school, at lunch time before I ate, before the evening meal and one last time before I fell asleep at night. As I grew older, I recognized a pattern in my prayer life. I was spending the majority of this time begging. Praise and thankfulness were expressed in the same repetitive phrases. I knew that this was wrong, but it was so ingrained in me to pray all the time that I just couldn't imagine doing things in a different way. Real change started for me when I came to a period of my life where I was being bullied and tormented. At first, my response was to feel sorry for myself. Then I crept ever closer toward depression and feelings of hopelessness. A friend of mine, who happens to be a doctor of psychology, noticed my state of mind and commented to me, "You do realize that tribulation is mandatory and suffering is optional? Right?"

Mind blown. I did not have to suffer. Suffering was a choice. From there on out I left the suffering, sorrow and misery behind. I decided that I would only pray when I had songs of praise to sing. I would be thankful for everything with which I had been blessed. I chose to look at what I had instead of whine about how hard my life was. I journaled volumes. I poured out my experiences onto paper and slowly I began to experience life differently. During this time I began learning to meditate which I found to be a very useful tool for breaking negative thinking and to improve mood and to reduce and finally eliminate panic attacks. Since the training I received was spread out over months and years without any feedback of which to speak, my meditative methods may be of my own creation rather than traditional. I have spoken to a few people about it, and they seem to feel that what I'm doing is pretty much on the mark.

In case it would be of help to others, I will describe my method as best I can.

I find a comfortable chair or lay flat on my bed. I get as relaxed as I can, breathing in through my nose and exhaling slowing through my mouth. When I first started, it took me about ten breaths to feel like I was ready, but now three deep breaths is plenty.

When I'm ready, I visualize roots growing out of the bottoms of my feet or back (depending on whether I'm sitting or lying down). I grow the roots deep into the earth and then begin absorbing energy from the earth around me. (This visualization grounds me to life and Earth.) As I absorb energy, I create a long hollow tube that I attach to my tailbone and I send the tube into the center of the Earth as a waste disposal system. Then using the energy that I have drawn into me, I cleanse my body of all negative energy and pain and use the waste tube to get rid of all the bad to burn up in the heat of Earth's core. Sometimes this is a very fast process and other times I need to spend longer to feel fresh and whole.

The next step is to reach out into the Universe and gather all that I want to draw into my life. I gather physical energy, laughter, joy, love, peace, good health, faithful friends, warm experiences, and courage. I let these things pour down over me like a pleasant warm rain that rejuvenates me with hope and strength. Sometimes I stop at this point. I've found that this process takes only a few minutes in my mind, but when I open my eyes to reality as much as forty five minutes have gone by to as little as twenty minutes.

If time permits, I visit a place of refuge that I have built in my mind. There is a place there to work through the sad and difficult things that have happened in my life. I keep the records of these events in the library of this house. I find it easier to deal with negative events if I can face them on my own terms and then close the book on the subject putting them on a shelf. I speak with my departed mother and other relatives in the living room of this house and I find peace in communion with angelic beings in that place. I'm in control of what happens there. It is a safe and secure place where none can enter without my permission. The place you create should be what you think of as beautiful, and the rooms you create need to reflect your own needs for healing as well as joyful celebration. You can also study here and ponder the miracles of the Universe. Its adaptations are completely up to you depending on what you need.

As you create this house, you will imagine a lock on the doors that only you can open. The interior of the house can be as detailed as you want. You may choose rugs, wall art, furniture or you may choose to furnish it over time adding rooms and decorations as you need them. This place is yours and yours alone. It should reflect you, your personality and your strengths. Don't be influenced by others. This is your sanctuary and your most private place.

There are many other constructs, probably a limitless number. If you are interested, what I've just told you about is a place to begin. Find a quiet place to begin your practice. Try to be consistent and choose a time of day that works for you. Practice for a half hour every day. It's not much time for a big benefit.

Prayer and Meditation

Martha was so confused. Robert was so shy and handsome while his twin brother had actually agreed to attend church services with her. Albert was charming and funny. He always made her laugh, but Robert had been her boyfriend for almost a month now. Both men were very good looking; tall, slender with broad shoulders and athletic. Al always had a twinkle in his eye and he was a war hero! Robert was shy and sometimes stuttered when he was nervous. She would base her decision on faith. She would choose the one who had agreed to come to church with her.

And so, after a whirlwind romance and one fist fight between the twins, Martha and Albert were married. Martha introduced Robert to one of her girlfriends and eventually they married as well.

Albert and Martha were off to a great start. There was only one problem. They had been trying to get pregnant for five years now and still there was no child. Robert and Margie hadn't been married as long and they already had a beautiful baby boy. Martha's heart ached for a girl child with black hair, blue eyes and all the gentle charm of Snow White. She prayed about it every day, but still, no child.

Martha decided that she and Albert should go to one of the church elders and ask for intercession with their problem.

It was a cold and blustery March day with only the knowledge that spring would come eventually to warm the day. The house they were approaching was grand. Uncle Deal was a successful baker who was also an elder in the church. Barette Bread was the best in town and Uncle Deal was a man who had tremendous will to get things done. If anyone's prayer could help them, his would be the one.

When he heard that they wanted to talk to him, he had invited them to dinner. The young couple opened their hearts to him and in so doing touched his heart. Uncle Deal said, "Why, Martha, of course, you can have a baby!"

After dinner they went to the library for privacy and there the three of them knelt down together in the center of the room and prayed. All were in tears by the conclusion of those heartfelt moments.

Martha always said that the next nine months were the happiest days of her life. It was Christmas Day at dinner when she felt the first of the labor pains. At five minutes to midnight a robust dark haired little girl with blue eyes was born to them. They named her Terra.

Prayer and Meditation

Recommended Reading List

- Jesus, Interrupted: Revealing the Hidden Contradictions in the Bible (and Why We Don't Know About Them), Bart D. Ehrman
- God's Problem: How the Bible Fails to Answer Our Most Important Question – Why We Suffer, Bart D. Ehrman
- Misquoting Jesus: The Story Behind Who Changed the Bible and Why, Bart D. Ehrman
- The Lost Gospel of Judas Iscariot; Betrayer and Betrayed Reconsidered, Bart D. Ehrman
- Peter, Paul, and Mary Magdalene: The Followers of Jesus in History and Legend, Bart D. Ehrman
- A Brief Introduction to the New Testament, Bart D. Ehrman
- Lost Christianities: The Battle for Scripture and the Faiths
- We Never Knew, Bart D. Ehrman
- Lost Scriptures: Books That Did Not Become the New Testament, Bart D. Ehrman
- The Apostolic Fathers, Bart D. Ehrman
- Jesus: Apocalyptic Prophet of the New Millennium Bart D. Ehrman
- After the New Testament: A Reader in Early Christianity, Bart D. Ehrman
- The New Testament and Other Early Christian Writings: A Reader, Bart D. Ehrman
- The New Testament: A Historical Introduction to the Early Christian Writings, Bart D. Ehrman
- The Orthodox Corruption of Scripture: The Effect of Early Christological, Bart D. Ehrman
- Controversies on the Test of the New Testament, Bart D. Ehrman
- The God We Never Knew, Marcus J. Borg
- The Gospel of Thomas, John Dart
- The Gospel of Mary of Magdala by Karen L. King
- The Gnostic Gospels, Elaine Pagels
- Adam, Eve, and the Serpent, Elaine Pagels
- The Origin of Satan, Elaine Pagels
- Secrets of Mary Magdalene, Dan Burstein and Arne J. De Keijzer